Following the sun

Following the sun

The pioneering years of solar energy research at
The Australian National University: 1970–2005

Robin Tennant-Wood

Australian
National
University

E PRESS

ANU
E PRESS

Published by ANU E Press
The Australian National University
Canberra ACT 0200, Australia
Email: anuepress@anu.edu.au
This title is also available online at http://epress.anu.edu.au

---

National Library of Australia Cataloguing-in-Publication entry

Title: Following the sun : the pioneering years of solar energy
            research at the Australian National University, 1970-2005
            / Robin Tennant-Wood.

ISBN: 9781922144126 (pbk.) 9781922144133 (ebook)

Notes: Includes bibliographical references.

Subjects: Australian National University.
            Solar energy--Research--Australian Capital
            Territory--Canberra--History.
            Solar energy--Research--Australian Capital
            Territory--Canberra.
            Research--Australian Capital Territory--Canberra.

Notes: Includes bibliographical references.

Dewey Number: 621.47

Cover design and layout by ANU E Press

Contents

Acknowledgements

Following the light of the sun, we left the Old World.

Christopher Columbus

This history would not have been possible without the assistance of, firstly, the solar energy researchers, and those directly associated with their work, who took the time to share their colourful stories: Stephen Kaneff, Peter Carden, John Morphett, Roger Gammon, Andrew Blakers, Glen Johnston, Andreas Luzzi, Ray Dicker, Martin Green and the late Keith Garzoli and Bob Whelan all contributed their recollections and opinions generously and candidly.

The assistance of the Centre for Sustainable Energy Systems at The Australian National University (ANU), through its director, Andrew Blakers, and former centre manager, Ray Prowse, was key to project. Andrew Blakers' patience and diligence in ensuring that the publication accurately reflects the University's solar energy research, his openness in providing access to all necessary support and the backing of all others concerned is gratefully acknowledged.

Also acknowledged with thanks are Chris Dalitz, from Country Energy, who contributed photos and current information about the White Cliffs solar energy station; Martin Green, David Mills and Monica Oliphant for their assistance in placing the story of solar energy research at ANU in a broader context; and Caroline Le Couteur, who recalled her work in the Commonwealth Department of Primary Industries and Energy. The library at the Australian Academy of Science generously allowed me to spend many hours on site sifting through the papers of the late Sir Ernest Titterton, which provided a good deal of insight into the workings of the Research School of Physical Sciences in an age before computers became the primary means of communication and storage.

Thanks go to Heather Neilson, Kim Wells and Artur Zawadski for reading and editing the manuscript.

Compiling the history of research in a single discipline at a university over a period of some 30 years, and incorporating the work and personal memories of a range of characters, has been a fascinating journey. These researchers were pioneers in every sense of the word: breaking new ground, challenging the status quo and working against the odds. They were, and many remain, world leaders in the field of solar energy. I am thankful to have had the opportunity to get to know them and to record their collective history.

Abbreviations

ABARE	Australian Bureau of Agriculture and Resource Economics
ABARES	Australian Bureau of Agriculture and Resource Economics and Sciences
ABS	Australian Bureau of Statistics
AFASE	Association for Applied Solar Energy
AINSE	Australian Institute of Nuclear Science and Engineering
ANSTO	Australian Nuclear Science and Technology Organisation
ANU	The Australian National University
ANZSES	Australian and New Zealand Solar Energy Society
ARC	Australian Research Council
ATICCA	Australian Tertiary Institutions Commercial Companies Association
COP3	Conference of the Parties
CSES	Centre for Sustainable Energy Systems
CSIRO	Commonwealth Scientific and Industrial Research Organisation
EGRET	Expert Group on Renewable Energy Technologies
ERDC	Energy Research and Development Corporation
GOML	glass-on-metal-laminate
IEA	International Energy Agency
ISES	International Solar Energy Society
MRET	Mandatory Renewable Energy Target
NEAC	National Energy Advisory Committee
NEI	Nuclear Energy Institute
NERDDC	National Energy Research, Development and Demonstration Corporation
OPEC	Organization of the Petroleum Exporting Countries
PV	photovoltaics
RSPhysS	Research School of Physical Sciences
RSPhysSE	Research School of Physical Sciences and Engineering
SG	Solar Generator
Sydney	University of Sydney
UNCED	United Nations Conference on Environment and Development
UNFCCC	United Nations Framework Convention on Climate Change
UNSW	University of New South Wales

Introduction

Solar energy in Australia is rapidly gaining in profile and acceptance at all levels of society, science, technology and politics. Increasing awareness of both the environmental issues associated with fossil fuels and the declining cost of renewable energy and its technologies, coupled with government schemes to encourage uptake of domestic solar power systems, has seen the sector grow very rapidly in recent years. Behind the government policy and climate change debate, the marketing and commercialisation and solar energy's coming of age as an alternative to conventional power, lies a story that goes back to the early 1970s. Commercially viable solar energy technology has not appeared *ex nihilo*. Rather, it is the result of decades of work, challenges, big projects and small advances, acceptance and rejection.

This book is the story of what preceded the current period of solar energy research and development. The story focuses on the solar energy program at The Australian National University (ANU), which began in 1970. Central to the story is the struggle to gain acceptance of the field of research by a university whose purpose, under Commonwealth legislation, was pure scientific research, and the difficulty of challenging the dominance of fossil fuels in the energy industry. It is also a tribute to those researchers who put their careers on the line in order to follow a belief that they could influence the course of energy research and provide a viable alternative to oil and other fossil fuels.

The history begins with the first major flagship project undertaken in solar energy research in Australia: the solar power station developed and implemented under the direction of ANU Professor Stephen Kaneff, in the remote western New South Wales mining town of White Cliffs. This project demonstrates, in many ways, the blend of scientific research, technical skill and entrepreneurialism that has characterised solar energy research, not only at ANU but Australia-wide, since the early 1970s.

Chapter 2 examines why 1970 was the right time for the development of a solar energy research group at ANU. Against the tide of energy research, which was moving towards nuclear energy, solar energy ultimately became a standalone research entity, challenging not only the energy research status quo, but also the image and perception of ANU as an academic research institution that eschewed applied research. Chapter 3 focuses on another flagship project for ANU researchers and the last undertaken by Stephen Kaneff: the Big Dish. Chapter 4 charts the development of ANU itself as a research university and how the emergence of the 'applied' science of solar energy was accommodated within it. Chapter 5 broadens the focus by looking at the role played by solar energy researchers in Australia, going back to the 1950s and the solar energy pioneers working at the Commonwealth Scientific and Industrial Research Organisation (CSIRO) and those who later worked at the University of New South Wales and The University of Sydney, competing with each other for the limited funding available but ultimately collaborating to advance public and political awareness of their field. This chapter also introduces the Australian and New Zealand Solar Energy Society and its role in solar energy research. Chapter 6 again returns to ANU and looks at the expansion and commercialisation of solar energy by the University.

At the time of writing, the field of solar energy is undergoing very rapid growth as the effects of climate change become evident and sustainable energy planning becomes central to governments' agendas. Chapter 7, therefore, deals with solar energy in the light of the understanding of climate change and environment policy, and how this has changed since the 1970s. Chapter 8 continues this theme and considers solar energy in a time of changing priorities and attitudes.

Over the years there have been many prominent and successful researchers who have worked at ANU in solar energy research, including Andrew Blakers, Kylie Catchpole, Andres Cuevas, Mike Dennis, Vernie Everett, Evan Franklin, Glen Johnston, Keith Lovegrove, Andreas Luzzi, Dan Macdonald and Klaus Weber. This account, however, deals mainly with the early pioneers in the field who worked from 1970 until the mid 1990s. The hard yards of the 1970s and 1980s, undertaken by such people as Stephen Kaneff, Peter Carden, Ken Inall and Bob Whelan, to name just a few whose work is documented in these pages, set the scene for later researchers.

As Australia's leading university, it is fitting that ANU has been a leader in energy research of a kind that will move us towards a sustainable 21st century, and beyond.

1. White Cliffs: From laboratory to reality

When viewed on the satellite imagery of Google Earth, the barren pockmarked landscape of White Cliffs in north-west New South Wales bears a startling resemblance to photographs of the surface of Mars. Most homes are underground. Except for the few roads and a small cluster of above-ground buildings denoting human habitation, the pitted mullock heaps, disused opal dugouts, and the copper-red vastness of the surrounding landscape could almost be images of the red planet. On the southern edge of the tiny township, a curious V-shaped structure stands in contrast to the rectangular regularity of the town's roofs. At a greater magnification, what emerges is the array of 14 paraboloidal dishes that was arguably the first commercial solar power station in the world. This tiny spot in one of the most remote places in Australia marks a milestone in international science and engineering, and a link between an obscure opal mining outpost and a nation's leading university.

The station was the brainchild of Emeritus Professor Stephen Kaneff, then head of the Department of Engineering Physics, Research School of Physical Sciences (RSPhyS) at The Australian National University. The funding came from the New South Wales government and Kaneff was ably supported by members of his Department including Head Technical Officer Robert Whelan, Mechanical Engineer Keith Thomas, Technical Officer Peter Cantor as well as Senior Fellow, Dr Peter Carden, who had led the preceding research. The White Cliffs solar power station was intended to be a demonstration of a 'commercial' solar power station as well as an experimental work horse. In retrospect it represents far more in terms of the advances made in a field which even now, and with widespread acceptance of the global need to reduce greenhouse emissions, is struggling to gain ground in an industry that can be as dirty as the coal that dominates it. The energy stakes are high and competition fierce. Australia's energy market is small by global standards but nonetheless voracious and the historical dependence on coal-fired power with its lobbying muscle, together with frequent challenges by the nuclear industry, makes entry to the market by the renewable energy industry difficult at best. This study documents the development of solar energy research at The Australian National University (ANU) from its modest but optimistic beginning in 1971 through to the turn of the century. The solar thermal power station at White Cliffs set a benchmark in solar energy and stands as a testament to the people who made it possible.

White Cliffs: The breakthrough project

In 1979 the New South Wales government was in electoral peril. Floundering in the polls against a public backlash, it needed a high profile project to take to the electorate. Premier Neville Wran believed that environmental issues held broad electoral appeal (Warhurst and Parkin, 2000: 295) so the prospect of a high-profile environmental project to take to the electorate would have been very attractive to a government in trouble. At the same time, Stephen Kaneff, always seeking financial support, had decided to undertake a major project in solar energy. The oil crisis of the 1970s had generated widespread public interest in alternative energy and the popular media had followed the research being undertaken in this area at ANU. Kaneff submitted a proposal to the Premier and subsequent negotiations held between these two and the Vice Chancellor, which were necessarily confidential, culminated in the dramatic surprise announcement of a grant of $800,000 for a solar power station to be built in a remote location in New South Wales. The original location proposed by the researchers was Fowlers Gap, where a remote research facility was already in place, but the Premier's press secretary preferred an even more remote location.

The township of White Cliffs, some 250 kilometres north-east of Broken Hill and 100 kilometres north of Wilcannia, at longitude 143°05′ east and latitude 30°51′ south, was chosen as the site for the prototype solar power station. The climatic conditions there are harsh: the town holding the Australian record for the most extreme range of temperatures. Summer temperatures are regularly in the 40s, often climbing into the 50s, and rainfall is very low at around 245 mm per annum. Population varies as it largely comprises itinerant miners and opal fossickers, but the 1986 census put it at 207 people in 102 occupied dwellings (ABS: 1988). The fact that the census also recorded 36 unoccupied dwellings is probably indicative of the transient nature of the population. Unconnected to the grid, the town was powered by individual private generators. Kaneff's proposal was a project that would undoubtedly generate wide public appeal for the government as well as test the research and emerging technology for the Department of Engineering Physics in RSPhyS.

The Premier's agreement to fund the project immediately caused some problems for the university in terms of how to deal with the offer. At that time, ANU was inexperienced in dealing with external funding for projects, and rather cautious about the strings that may be attached to money derived from outside the research community. For a research university in the tradition of a liberal-arts college, the acceptance of external project funding could pose a threat to scientific independence and integrity, or be perceived to do so. Kaneff recalls that the only means whereby external funding could be directed to a project such as this was via the Vice-Chancellor. So Premier Wran accordingly wrote to Vice-Chancellor Professor Anthony Lowe and made the offer of $800,000 – roughly equivalent to $3 million in 2012 dollars - for the development of a solar thermal power station at White Cliffs.

While the NSW government offer of funding in response to Kaneff's proposal for the project was unprecedented, the amount on offer and the prominence of the project itself made it difficult to refuse. For the university it represented more than a large injection of project funds. Until this time the university had been less than supportive of the Department of Engineering Physics in their pursuit of solar energy research. Cuts in funding to solar projects had, on occasions, prompted negative media coverage and subsequent public outcries. The offer by the New South Wales government was an opportunity for ANU to take up a high profile project with wide public appeal as an indication of support for research and development in an increasingly accepted and popular field. Whilst the reaction of the Vice-Chancellor upon receipt of the Premier's letter is unknown, the figure of $800,000, the source of the funding, and the proposed project itself, would presumably have prompted more than a ripple in the Chancelry of a university with a very cautious and conservative approach towards external funding. Regardless of any initial vice-chancellorial misgivings, however, in July 1979 ANU was formally commissioned by the New South Wales government to construct a solar thermal power station with commercial application.

The purpose of the White Cliffs solar power station was to "ascertain the feasibility and potential (both technological and economic) for providing electric power in conditions which exist over much of inland, remote areas and off-grid Australia" (Kaneff 1991:5). The intention was to construct an experimental system to operate for two years. In fact, it operated for 11 years with the assistance of locals, Peter Thompson and later Bill Finney, who attended to it on an 'as required' basis. As a feasibility study into the functionality of solar thermal systems for commercial domestic power supply, it was a success fulfilling many of the expectations of the local community, the researchers and the NSW government.

Problems of location

The whole point of choosing White Cliffs as a site for the project was its remoteness, but this was a double-edged sword. As a sparsely populated opal-mining town on the edge of the desert and not connected to the national energy grid, White Cliffs met the requirements of both the government, in taking a high-profile, future-looking and environmentally sound project to the electorate, and the research team, in providing an off-grid town with commercial solar power. The remote location that fitted the project plan, however, also presented problems to the solar energy research team. John Morphett, the founding manager of ANUTECH and former laboratory manager for the Research School of Physical Sciences, describes the location as 'diabolical'. The sheer distance involved posed as many difficulties as did the new technology itself. Fowlers Gap, which was the original suggestion for a location, was almost as remote but had an existing research facility which would have made the project logistically more tractable, but White Cliffs had nothing. Everything had to be either built on-site or built in Canberra and transported over 1100 kilometres, some of it over an unsealed road. Because of the lack of facilities at the White Cliffs site, a building in the Fyshwick industrial area of Canberra was rented and the concentrating solar collectors for the power station were built there, with many of the components for the overall system constructed in the Research School and Engineering Physics workshops and by Canberra companies.

The station served as an experimental unit allowing the scientists and engineers to design, develop and test appropriate technologies for what Kaneff describes as "a range of solar stations". Added to the problem of the location was the requirement that the project had to provide "power reliably and continuously on a stand alone basis (with diesel backup); that is, [it] had to progress from conception to useful effective operation in one step" (Kaneff, 1991: xxi). While meteorological records for the area indicated about 3000 hours of sunshine per annum, providing very good levels of insolation, meteorological records of wind conditions were sparse. There were anecdotal reports of extreme storms and wind, including stories of cars being sandblasted clean of their paint during dust storms, which needed to be addressed in the design of the solar station. The low rate of annual precipitation did not pose any difficulties, but the potential for dust storms, particularly the incidence of the fine aeolian dusts that can reduce insolation and soil collectors, had to be considered.

The field station established at White Cliffs was a base for research. It became a valuable resource for researchers from other universities requiring a remote and climatically appropriate location, including as a site for wind energy research. Kaneff himself undertook over 100 trips over the first seven years of the project, each being a round trip by car of 2,200 kilometres. He reported in 1991 that this travel afforded an opportunity to "reflect on our intrusion into this sensitive area" (Kaneff, 1991: xxii), specifically the abundance of wildlife long since gone from settled areas.

Building the first solar power station from scratch

Once the site had been chosen, Peter Cantor, a senior technician in the Department of Engineering Physics, spent time at White Cliffs surveying the site for the station and procuring the necessary equipment, including crane equipment, for the placement of the solar collector dishes. Kaneff rented a workshop in Fyshwick, and much of the first year was devoted to "research and development on several concepts for collectors and other components" (Kaneff, 1991: xxi). It was not until August 1980 that a final decision was made on the actual configuration and systems that would be used at White Cliffs.

The key, according to Kaneff (1991), is simplicity. He compares solar and nuclear energy by way of illustration. Whereas the production of nuclear fission energy requires highly complex technology and produces a complex waste, solar energy is perceived by many to be relatively simple. The concept of harnessing the nuclear energy of the sun to produce electricity is also a far easier one to grasp than the use of a finite resource, uranium, to create fission energy through a process that is very costly in resources, money and time. Solar energy produces no hazardous waste or threat of radiation poisoning. As Kaneff succinctly points out, the only radiation danger is sunburn.

During the early years of the eight year period prior to the White Cliffs grant, Kaneff's colleague Peter Carden set an agenda for solar energy research to meet three criteria. Those were that it should be complementary to other Australian research, should fit as well as possible the 'publish or perish' culture of the Research School and should aim squarely at addressing the core problem of large scale utilisation of solar power. Research into the general economic viability of solar power highlighted the importance of recognising the dispersed nature of solar radiation and the inevitable need to provide large areas of some sort of intercepting medium. This turned out to be critical in any cost analysis and set severe limits to the cost per square meter of whatever intercepted the insolation, leading to the concept of employing thin metal paraboloids press-formed in the way car bodies are shaped. Clearly the car industry had the expertise to mass produce such dishes to the required precision and finish and could help answer the question of probable cost.

During this period attention was also given the question of steering collectors so as in the direction of the sun. The idea of computer control incorporating a certain degree of artificial intelligence was developed. If for example each dish had the ability to set its own coordinate system then a considerable savings could be had by avoiding the need to precisely locate and set up each dish and even avoid the need for firm concrete foundations. All these concepts were developed and embodied in peer reviewed publications. Peter Carden and Bob Whelan visited the University's observatory at Siding Springs to investigate the feasiblity of using the technology developed for telesceope mirrors for the solar collector dishes.

Another problem emerged in the form of inexpensive transport of heat from hundreds of dish collectors. Thermo chemical processes seemed worth exploring and Carden settled on the ammonia system as one needing evaluation. The concept involved feeding ammonia liquid to the focus where heat was absorbed in its dissociation to nitrogen and hydrogen. Each dish was to be fitted with a reaction vessel and counterflow heat exchanger that enabled the fluids to efficiently transit between ambient and focal temperatures. The cooled gases could then be piped to a central recovery plant carrying energy cheaply and without loss. This idea promised a substantial cost saving because it obviated the need for kilometers of thermally insulated pipes. Recovery of the heat energy occurred at a central plant when ammonia was re synthesised.

The experimental nature of the project earned its stripes again with the development of another line of research: the feasibility of storing hydrogen in aquifers. All this work required additions to Peter Carden's existing Energy Conversion Group: Dr Owen Williams, and PhD students Lincoln Patterson and Kieth Lovegrove. Most of the work was necessarily theoretical in the first instance and was well documented in over fourteen peer reviewed journals as well as presentations at Australian and overseas conferences.

Carden succeeded in obtaining Commonwealth grants for the thermo chemical work and also to test the idea of forming paraboloids from aluminium sheet. Small models indicated that the sheet could be sucked into a mould but there was no guarantee as to the accuracy of the first trial. Yet if such a formed dish could be made it could allow many valuable engineering and field tests. However, after several attempts it became evident that there were problems with the size of the shell in that the surface area was larger than the available metal sheeting width and required welding sheets together. During the forming process, the dishes consistently cracked along the line of the weld. There was hope that the dishes for White Cliffs could be made this way but there was always little chance for this. This research had a long way to go and still has.

Another concept developed by the Energy Coversion Group was a facility for pairs of dishes to close up, like clamshells, in extreme weather conditions to protect the dish integrity. Unfortunately, the time required to finetune this technique did not fit into the timeframe required for White Cliffs.

The White Cliffs power station consists of 14 paraboloidal dishes, each covered with approximately 2000 small mirror segments to allow the mirror surface to conform to the dish shape. Each dish has an area of about 20 square metres. Denied any outcome of previous reseach, Kaneff and his team went back to the drawing board and designed paraboloids constructed from fibreglass cast on a plaster-of-paris mould in the Fyshwick workshop. This design met the criterion of being inexpensive and also allowed the technicians to establish the optimal shape. Once the dishes were cast, students were employed at a (then) generous rate of about nine dollars an hour to undertake the tedious task of fitting the mirrors, the glass for which was supplied by Pilkington and cut locally. Eighteen of the dishes were constructed in the Fyshwick workshop. Of the four not used for White Cliffs, one - a prototype - was eventually sent to Maryborough (Victoria); two were sent to feature in an exhibition at Knoxville, USA; and one was retained at ANU and is still used for solar driven thermochemical experiments.

The final size of the 5 metre diameter dishes was largely determined by considerations of safety in ensuring survival in extreme winds, and especially the need to transport the units to the site. Once in place, the paraboloids could track the sun. The 14 dishes were installed to operate as separate modular units, centrally connected but with each unit able to operate independently even in the event of failure of communication with the central control unit. The system was fitted with a wind monitor and the dishes were pre-programmed to park in a horizontal orientation, facing the sky, when wind speed exceeded 80km per hour.

Solar thermal heat collected by the paraboloids heated water to produce high temperature, high pressure steam, which in turn powered the engine to drive the generator, thus producing electricity. Straightforward engineering, but both Kaneff and Whelan recalled that the building of the steam engine was the most troublesome part of the enterprise. Kaneff (1991:4) reported that "the major part of the project time, effort and other resources was required to establish the steam engine … and water/oil treatment systems as robust, reliable working units". A Rankine Cycle Uniflow Steam Engine with an output of 25kW was originally proposed. However, this was not available commercially, and a retired naval engineer officer and steam car enthusiast, Commander Graham Vagg, was approached to join the team. Vagg had developed a technique for converting a four cylinder Volkswagen to run on steam and suggested the idea that a three cylinder Lister diesel engine could similarly be converted to run on steam. The advantage of the Lister engine was that each cylinder and head is removable. In engaging Vagg to build the steam engine, however, what the research team did not know, according to Kaneff, is that if a steam car enthusiast builds a steam engine for a car, "he will shout with joy when the car can go round the block and come back without breaking down". While Vagg's engine was conceptually good, it was also highly unreliable. Whelan recalls that: "We had piston failure that often led to catastrophic damage to the engine as well. I ended up changing it so that instead of cutting the rig-welded pistons I took a different approach and modified the top of the piston." By this stage Dr Ken Inall, a research engineer who had previously been working on wind energy, had joined the team. He and Whelan worked on perfecting the steam engine for the solar power station. This took up a good deal of 1982.

The establishment of ANUTECH

Because the funding arrangements for the White Cliffs project were unprecedented, the university had no mechanism for managing such a large project utilising external funds. Other universities had commercial divisions to manage comparable projects of a commercial or technical nature, and it was decided that ANU should establish a similar organisation to manage White Cliffs. The Deputy Vice-Chancellor, Professor Ian Ross, was instrumental in the establishment of ANUTECH in 1979, and Mr John Morphett was appointed as founding manager with the responsibility to set up the organisation. All business with the NSW government relating to the White Cliffs project was conducted through the company, thus removing the need for the university to be directly involved with the commercial operation.

Morphett recalls that the White Cliffs project "scared the hell" out of the university because it realised that it hadn't really gone far in supporting solar energy research. ANUTECH was formed as a means of protecting the university in the event that the NSW government did not honour its part of the contract. Unlike most university research grants, this contract was perceived to be commercial in application and outcome – this was new territory for a university hitherto based primarily on academic research. It was further considered by the university that a company could administer commercial operations in a much more efficient manner than a university administration.

It was subsequently decided that there should be a commercial component to the project and a commercial company was sought by ANUTECH to become involved. A company called Environ came on board and Ken Fulton, a chemical engineer who had formerly managed Environ's New Zealand operations, came to work on the project as the representative of the commercial partner. Morphett recalls that Fulton's involvement was not readily accepted by the researchers. Their views were still largely defined by what Morphett refers to as: "old university thinking ... (Sir Mark) Oliphant's attitude to life ... This department was Oliphant's old department." This resulted in a reluctance to accept Fulton as the outside representative of a commercial company, and instead of the project benefiting from Fulton's engineering experience, his role became that of project management. Fulton, being a competent and likeable person, was nonetheless able to work efficiently, if not to potential, with the project personnel.

The advantage of the project working through ANUTECH was that the entire project was managed centrally. All the construction, development and engineering for the White Cliffs power station was undertaken 'in house' with the solar energy team's own personnel together with whatever resources were at their disposal. Nothing was outsourced and there were no external tenders awarded. Specialised staff could be hired and external expertise could be obtained as required without needing to go to a formal recruitment process. This enabled project staff to do things "promptly and economically", according to Morphett. Quality and time management were enhanced by the system rather than reduced, allowing the solar energy team to get on with the task of building the power station without the interruptions that can often hamper such projects while bureaucratic process is followed. Rapid progress was desirable to meet with the electoral schedule of the funding body.

Success

The final working power station at White Cliffs is described by Kaneff:

> A large storage battery, the power from which was used to drive a shaft connected to an alternator which generated 50 Hz alternating current which in turn was connected to supply the town load …….. The system was also connected to a steam-driven engine powered by solar-generated steam via a 'free wheel' coupling, such that when there was solar energy available, the steam engine drove the system and supplied the town with electricity. Any solar energy left over was stored in the battery for use when the sun was not shining. If there was inadequate energy from the sun, the electricity generated came from both the sun and from the battery. If the battery became nearly discharged, the diesel engine cut in automatically to supply electricity to the town and would do so until further energy came from the sun. This arrangement meant that all connected customers always had power up to the limit of their allocation, and all solar-generated energy was used. Since the power supply was of very limited capacity [25 kilowatts], there had to be a limit imposed on each customer. (Kaneff, interview 2008)

In 2006 the White Cliffs solar power station, which first brought electricity to this settlement in 1981, was declared a National Engineering Heritage Site. Along with such nationally, indeed internationally, recognisable works as the Snowy Mountains Hydro-Electric Scheme and the Sydney Harbour Bridge, this comparatively unheralded enterprise is now formally acknowledged as a leading benchmark in engineering. Although decommissioned in 1994, the power station still stands and is now part of the town's tourism program.

2. Why solar? Why then?

Solar energy research began at ANU in 1971 at a time when renewable energy and ecological imperatives were not high on the social or political agendas. When Stephen Kaneff began his work in solar energy at ANU in 1971, Australia had a population of 12.5 million, the Prime Minister was Billy (later Sir William) McMahon and the national anthem was *God Save the Queen*. The average weekly wage for men was $95.60 and for women $73.60 (Australian Bureau of Statistics, 1971a), and there was only a 37% female participation rate in the labour force. A brand new Holden HQ sedan cost $2,370; the median price for a house was $21,200 in Sydney and $18,000 in Canberra (Abelson and Chung, 2004: 8). Australia still had a military presence in the Vietnam War and had no formal diplomatic or economic relationship with China. There were only 15 universities in Australia, with a total student enrolment of 123,776, of which 31% was female and 69% male. It was a socially and politically conservative period, dominated internationally by Cold War politics and domestically by some of the most colourful and controversial state premiers in history: Robert Askin, Henry Bolte, Joh Bjelke-Petersen and Don Dunstan. Major social discussions were emerging in the public domain including the women's equal pay movement, arguments over the Vietnam War and environmentalism. The national economy was stumbling at the end of the Long Boom period. In this social climate, the decision to pursue energy research in the largely uncharted area of solar energy must have seemed a somewhat radical departure from the expected path for an academic at The Australian National University.

Australia's need for electricity was rapidly growing due to the post-war economic boom. The Snowy Hydro Scheme was operational, but the ABS 1971 Yearbook noted that, with the exception of the Snowy Mountains, Tasmania, and "the narrow coastal strip along the east coast of the mainland", rainfall for the rest of the country was insufficient to make hydro power viable and that:

> By far the most important source of energy used in the production of electric power in Australia is coal. At 30 June 1970 thermal power equipment represented 70.7 per cent, hydro plant 26.9 per cent, and internal combustion and gas turbine equipment 2.4 per cent of the total installed generating capacity. (Australian Bureau of Statistics, 1971b: 947)

The Yearbook went on to predict that:

> The future electric power plants on the mainland of Australia will be predominantly thermal or thermo-nuclear installations, and in an electrical system in which the greater part of the energy is generated in thermal plants it is usually found that the hydro installations operate to the best advantage on peak load. (Australian Bureau of Statistics, 1971b: 951)

Why, then, were Kaneff and his colleagues moving away from the predicted trends of coal and nuclear power to follow the sun?

A road less traveled: The decision to follow the sun

For Kaneff, nuclear energy simply was not an option. He recalls that prior to 1970 the ANU Department of Engineering Physics was already involved in a number of projects investigating fusion energy at the small experimental scale. The department, which began as the Department of Particle Physics under Sir Mark Oliphant, was changed to Engineering Physics when one of the chief tools of particle physics, the homopolar generator, became available for high power, very large volume experiments, although these were still mainly directed towards the kind of work that had been undertaken in particle physics. There was a department of Nuclear Physics in the Research School but Kaneff's Department of Engineering Physics was looking at other aspects of fusion energy, including magnetic confinement fusion. There was an expectation among researchers and energy policy makers alike at that stage that nuclear energy was going to be very successful, safe, and would be produced inexpensively.

> Of course the people in the know knew that that wasn't the case, but that was the view that was being put out. Well, we didn't accept that view, we happened to know a bit more about the kind of things you get out of nuclear energy that we decided that was not the solution, certainly not nuclear fission energy. (Kaneff, interview 2008)

Nuclear fusion energy entails the fusing together of light elements such as hydrogen and tritium (obtained through a supplementary reaction from lithium) rather than the fission of heavy elements such as uranium. Fusion was being promoted as an even better long term energy option than nuclear fission. Kaneff held real concerns regarding nuclear energy. Many of his concerns arose due to the lack of time, and therefore research, in nuclear energy between the first laboratory nuclear chain reaction, the dropping of nuclear fission bombs on Hiroshima and Nagasaki, and the call for nuclear power-generated warships, submarines and electricity generation.

> As far as fusion energy was concerned the first fusion bomb was dropped on Bikini Atoll in 1951 … and here it is, more than 50 years later, and there is still no controlled thermo-nuclear fusion. What that means is that it is very difficult to achieve, if ever, in the laboratory. It's what the sun gives us – power. We're using fusion energy but it comes from the sun – a very safe place to be because we don't have to deal with the waste, and the only radiation we have is sunburn. (Kaneff, interview 2008)

Peter Carden, held similar misgivings about nuclear energy, particularly the difficult engineering problems associated with nuclear fusion power generation. Carden had already gained international recognition for his work in high magnetic field generation, and the establishment of a high field magnet laboratory in the RSPhysS. He had also experience in superconductivity research at MIT. Given that nuclear fusion reactors relied on magnetic confinement he was uniquely placed to study fusion reactors from an engineering point of view. He was able to do this during a period of study leave at Oxford University. At this time Carden was convinced that the new cutting edge of research was going to be long term sustainable energy sources. If the answer was nuclear fusion then he knew he could make a contribution. But was nuclear fusion the answer? It took a few months for him to decide that the engineering problems inherent in a commercial fusion power source were going to be almost insurmountable. With that option disposed of he turned to what he instinctively knew was the only plausible long term energy option for the world, solar. In December, 1970, towards the end of Carden's period at Oxford, he met with Kaneff in the farm house where he resided near Oxford and during a long discussion he gave his assessment of fusion and belief that somehow solar had to be the answer. By the time he returned to Canberra Carden had made up his mind being now determined to take up solar energy research. This accorded well with Kaneff's plans for the department as he wished to introduce this field but needed a high level researcher to help start it. Carden, he felt, would be an ideal candidate for this role. He subsequently formally introduced the field with the objective to ascertain the feasibility of solar energy as a source for providing substantial new benign energy resources and with an open charter as to what to do. Thus was formed the new Group consisting initially of just one person: Peter Carden. As he developed his agenda and new people were added he headed what was to become the "The Energy Conversion Group" within the Department of Engineering Physics with objectives to ascertain the Feasibility of the mass utilisation of solar energy.

Carden and his collegues were soon to be encouraged (interview, 2008) and influenced in the early 1970s by the famous book *The Limits to Growth*.

The Arab oil crisis occurred about 12 months later, giving the research added momentum, and Carden believes this was also the point at which the public became interested in the solar energy research being carried out, in much the same way as current interest in climate change is again focusing the public interest on renewable energy, and in particular solar energy.

By the end of 1975, Peter Carden's Energy Conversion Group included, Owen Williams, Winston Revie, Bob Whelan and five other staff. Williams helped exclusively with solar-driven thermochemical systems while Carden pursued the quest for economical solar collectors. He had ascertained that sheet metal reflective paraboloids had a chance of meeting the strict economic limits imposed by the competition from coal and oil and sought methods for mass producing these and controlling large numbers with computer technology.

Whelan remembered strong public interest, although much of it, he believes, was driven by media 'stunts' of the type in which demonstrations of solar energy for the cameras included using the solar dish foci to ignite pieces of wood or melt stainless steel. Whelan recalled that, while this was spectacular and made for good colour photos in the newspapers, community interest in the actual technicalities of the research was minimal. There was, however, considerable support for their work evident in the public outcries whenever it was reported that funding for solar energy research had been cut. In a view shared by many of the other early solar energy researchers, Carden's explanation for the level of public interest is simple:

> [It was] because of the oil crisis. It was the same sort of public interest then as there has been just recently and it's such a shame that it didn't take off at that time and continue through to now. But at that time, although there was a great public interest, there were great forces against us. (Carden, interview 2008)

The 'great forces' mentioned by Carden included the coal and nuclear power lobbies, which gained considerable political ground during the oil crisis. In a 1974 interview for the NSW Institute of Technology newsletter, *Communique*, Dr Terry Sabine, the Head of the School of Physics and Materials, N.S.W. Institute of Technology, commented that the oil crisis had resulted in an increase in government funding for solar projects bringing it to $500,000 for 1975 compared to the $10m funding to be provided in the same year to the Atomic Energy Commission. It was clear where the power lay. (*Communique*, 1974: 4)

At ANU, community support notwithstanding, the Department of Engineering Physics was facing similar battles over funding research in solar energy. All political indications pointed to a future powered by coal or nuclear energy, and research into solar energy seemed to be a diversion at best into a technology that would never be part of the mainstream. In December 1974, the Head of the Research School of Physical Sciences, Professor (later Sir) Ernest Titterton, informed Stephen Kaneff that the Australian Institute of Nuclear Science and Engineering (AINSE) Council had approved funding for only three of the five solar energy projects proposed. Given the focus of AINSE is essentially on furthering research in nuclear energy, it was perhaps unsurprising when a year later, only one project actually received funding. In 1976-77 government funding to universities was tightened, resulting in the RSPhysS requirement that solar energy needed to operate on outside funds if it was to continue.

In the face of such opposition, others may have abandoned the idea of solar and turned instead to the more conventional areas of research in the energy field. However, Kaneff and his small but growing team, with a firm belief in the scientific integrity and practical application of their work, looked elsewhere for the necessary research funding. Between 1971 and 1985, the Solar Group published important scientific papers and reports. Carden's work in solar thermo-chemistry continued with funding until 1977 when university funds were discontinued. While the department could continue to pursue their solar work with departmental staff, they couldn't access additional university research funds. Because of the absence of Government funding mechanisms for applied science and engineering at universities at the time, the team turned to the private sector. Over 100 potential backers were contacted by Kaneff and Carden in 1977, with a grant for the thermo-chemical work finally coming from the unlikely source of Uncle Ben's pet food manufacturer, with a donation of $50,000. This company had a policy of donating money to a good cause every time they opened a new facility, and Kaneff's approach coincided with the opening of a new plant near Bathurst.

In 1978, Kaneff was contacted by the New South Wales government regarding ways in which solar energy usage could be enhanced. This contact came about because of the external focus and activities of the group as well as its reputation through published works and conference presentations in Australia and USA. Kaneff remembers that, while he didn't want to derail the thermochemical work, which was producing good results, the opportunity for leverage through the New South Wales government was too good to miss. The area of research which he considered needed further development at that time was solar collectors: "It so happened that when the NSW government asked us what we could do to enhance solar energy immediately, I was able to put up a proposal for a small solar thermal power station in outback NSW because they wanted remote areas and you couldn't get much more remote than that – in this part of the world anyway." In 1979 the project details were finalised and funding from the government agreed upon. The White Cliffs project was up and running.

The Oil Crisis and changes in international policy

By the early 1970s, there were indications that oil would not be in endless supply. Although what came to be known as the Oil Crisis occurred largely due to economic and political factors rather than the actual availability of oil resources, it sharply focused the attention of energy policy makers, economists, leaders of industry and consumers on the limitations of oil-dependent industrial expansion and unlimited consumption, based on the simple principle of supply and demand. Concerns about security of energy supply from a political point of view also emerged. In 1972 and 1973 there was a major surge in demand for oil, driven by a simultaneous surge in industrial growth in the US, Europe and Japan (Vernon, 1976: 3). Vernon states that the increased demand for oil occurred partly:

> … because of delays in bringing nuclear power plants into operation, and because of various antipollution controls [which in the US] reduced somewhat the efficiency of gasoline. (Vernon, 1976: 3)

The OPEC nations increased the price of oil fourfold, nationalised the oil producing assets of the international oil companies and threatened to disrupt the oil supply to developed nations. As a result of what became a complex interweaving of politics and industry, the international political landscape shifted permanently as the relationships between oil-exporting and oil-importing nations altered to facilitate and accommodate the new economic situation.

In 1975, with the political situation still volatile, the then US Secretary of State, Henry Kissinger, oversaw the establishment of the International Energy Agency (IEA). Based in Paris, the IEA's role initially was to facilitate a collective response to the oil crisis and any future disruption to global oil supply. Over time it has evolved as an international energy policy body as well as functioning as a non-partisan body for cooperative energy security. Around the same time as the IEA was established, the Ford Foundation Energy Policy Project in the USA released its final report, *A Time to Choose: America's Energy Future*, in which the authors proposed changes to existing energy policy including more efficient energy options and alternative energy sources that would maximise conservation of energy supply. The Carter administration accepted the report and began to provide federal funding to alternative energy research, including the establishment of the National Renewable Energy Laboratories (NREL).

In the 2006 documentary "A Crude Awakening", on the prospect of what is now popularly known as 'peak oil', the petroleum geologist Dr Colin Campbell asserted that the peak of oil discovery had occurred in the 1970s and that we were only then, some 35 years later, facing the peak and subsequent decline in oil production. As early as the 1950s, an American geophysicist, M. King Hubbert, predicted 'peak oil'. In a 1976 television interview, which was repeated in the 2006 documentary, he discussed the research underpinning Campbell's position: that oil discoveries peaked in the 1970s and production would peak in the 2000s before declining rapidly. Oil price volatility in the 1970s, combined with the growing environment movement (which incorporated the anti-nuclear movement) and US federal support for renewable energy, began to change the way the developed world regarded its energy sources. This was the global picture when Premier Wran provided funding for the White Cliffs solar thermal project.

The Club of Rome's influential report on human development, *The Limits to Growth*, published in 1972, added a further dimension to the question of energy. While noting that "the technology of controlled nuclear fission has already lifted the impending limit of fossil fuel resources [and that] it is also possible that the advent of fast breeder reactors and perhaps even fusion nuclear reactors will considerably extend the lifetime of fissionable fuels, such as uranium" (Meadows et al, 1972: 131), the authors were very cautious in their evaluation of a future nuclear-powered world on the basis that, with even the most optimistic outlook on resource availability, increased pollution and its associated costs would still pose limits.

> 'Unlimited' resources thus do not appear to be the key to sustaining growth in the world system. Apparently the economic impetus such resource availability provides must be accompanied by curbs on pollution if a collapse of the world system is to be avoided. (Meadows et al, 1972: 133)

Through a system of modelling, a world with 'unlimited energy resources' was promulgated, with the model showing that population growth, food production and industrial output would be impeded and then reversed by an exponential increase in pollution generated by the provision of unlimited energy. According to the 1972 model, this decline would occur in the first half of the 21st Century.

The Arab oil embargoes of the 1970s assisted in creating a public acceptance for solar energy that was reflected in increased funding and political acceptance. Dr Roger Gammon, who arrived at the New South Wales Institute of Technology from the UK in 1972 to undertake solar energy research and was later associated with ANU through the White Cliffs project, recalled that as a direct result of the oil crisis his department was able to advance their work significantly:

> Our total budget for one year was close to a million dollars. I know some of the projects right over the life of the funding. The evacuated tubular collectors were over a million dollars, so there was plenty of funding, a lot of interest in the media and the public, and I think we all felt it was very exciting times and renewable energy in solar energy systems were just around the corner...
>
> It was only the economics and the competition from cheap coal-fired power stations and cheap other fossil fuel energies that really killed it all off. Once the oil crisis of the 70s was over, the economics never stacked up and governments just abandoned the whole thing. (Gammon, interview 2008)

In a 1977 radio interview on a program called *Inside Sydney*, the interviewer, Peter Young, asked Gammon about public interest in solar energy. During the interview Gammon mentioned that an extension course on solar energy being run by the Institute had over 150 applicants – well over double the numbers they had expected. The extension course, consisting of eight evening lectures and some weekend demonstrations, was to cover such areas as industrial and domestic heating and cooling, solar energy and architecture, alternative technology and 'home handyman' solar technology, electricity power generation, and extensive computer simulation demonstrations to show "how Australia would benefit if it adopted a solar energy policy". The nature of solar energy is such that it appeals to the layperson in ways that other energy forms do not: it is difficult to imagine nuclear or thermal engineers running courses for home handyman enthusiasts to build their own backyard nuclear power plants. Solar energy, on the other hand, is accessible and explicable.

The sun is a vast, visible, low-pollution and inexhaustible source of energy that can be harvested without the need for huge, complicated installations. The enormous growth of roof-mounted photovoltaic systems in recent years is testament to this. Coal-fired and nuclear power stations come with innate environmental risks and represent a level of industry, security and technology that physically removes the source of power from its consumers. Thermal power plants are built away from urban areas while nuclear plants are isolated and hidden by fences and tight security systems. The implicit message is that these places are not safe to be near, nor are they aesthetically pleasing, and thus need to be sequestered from the human environment. Solar panels, on the other hand, cause minimal hazard to the public and the environment. Domestic units are clearly visible on many roofs in the suburbs and the larger commercial solar farms resemble larger versions of the panels anyone can have installed on their roof. These factors have, over the past four decades of solar energy research, contributed to its positive acceptance by the general public.

Vision and rear vision

According to Kaneff, the university's attitude towards the sort of applied research being done by the Department of Engineering Physics in solar energy was somewhat myopic. The prevailing conservatism of the Research School of Physical Sciences meant that the pursuit of "pure" academic research, as measured by traditional metrics such as journal publications, would take precedence over the applied research of the kind being undertaken by Kaneff's department. Applied research, as perceived by senior staff of ANU, was best undertaken by institutes of technology. This view was illustrated by the provision of funding to the NSW Institute of Technology's School of Physics and Materials to enable them to carry out their solar projects at the same time as Kaneff was being told that his funding was being withdrawn. Indeed, whilst

ANU has an excellent track record in research, research metrics show that it remains relatively weak compared with other "Great 8" Australian universities in applied science and engineering. Why is this?

An unwillingness to become involved in politics is one possible answer, although, because the university itself exists through an act of parliament, ANU has always been closely aligned with politics. However, the high level legislative politics and the diplomatic negotiations between Government and University Council for the establishment of the university in the 1950s and 60s bore little resemblance to the often grubby electoral politics of money and media that tends to go on now. Funding has usually hinged on political decisions and agendas.

Another reason is simply that, for a premier research institution, there was considerable pressure to maintain a high academic standard, as measured by journal publications and similar traditional metrics, at a time when institutes of technology existed to perform the function of applied research. It was deemed inappropriate for ANU to work in any field other than "pure" academic research. Kaneff and his team broke the mould, and once this was broken the opportunities to blend academic research with practical and commercial applications began to appear.

The direction taken in the 1970s is not without its critics, and even Peter Carden, leader of the research team at the time, believes that it was a mistake to attempt a demonstration project such as White Cliffs without having done all the research to support it. Carden was always conscious of the culture of the RSPhysS and tried to accommodate by making the economics of solar concepts the focus of the group's research. His considerable engineering experience with Sir Mark Oliphant's projects and high field magnets had presented many 'we could have worked that out before we started' occasions caused by a hasty 'suck it and see' approach. He was always fearful that White Cliffs would play into the hands of the coal lobby by demonstrating that solar power was uneconomical. As it happened it took extra work to fix problems as they arose in the glare of the political spotlight. Scrutiny was particularly intense because the White Cliffs funding was a political decision, and therefore an opportunity for the Opposition to attack the New South Wales government.

Kaneff now, with the benefit of hindsight, believes that the reason solar energy was not a high priority for the university establishment and administration actually had little to do with academic niceties and high academic standards. From his personal contact over the years with the main administrators concerned, along with colleagues and fellow scientists, he sees that a view prevailed, and even now prevails in some areas, that solar energy is trivial and cannot amount to much. The perception he has received is that if solar energy, being so 'simple', was worth researching, someone would have done it a long time ago.

> In reality it is so complicated that [critics of solar energy] do not envisage how to proceed with it; they do not have the vision to see and understand how it can really be the only long term benign means for a long term energy future. (Kaneff, comment 2010)

It is also true that had there not been a White Cliffs demonstration system at the time, it is not likely that there would have been much progress at all in solar energy at ANU, since outside funds were available for short term research and development but not for long term solar thermochemical research or further investigation into design and manufacture of paraboloidal mirror collectors. Gaining funds for White Cliffs was instrumental in enabling the survival of thermochemical research, not least by enhancing the public visibility of the group. Research approaches differ as well, and while some see that projects are almost never ready with complete information to implement in the real world, others believe that moving to the real world reveals problems which need solution and cannot be even predicted in academia. Kaneff considers that the White Cliffs project came along at just the right time to allow real progress to start, and as in all engineering progress, one rarely if ever has all the information before starting.

Notwithstanding the critics however, vindication, if any is needed, for Kaneff and his team, exists in the current interest and success of solar energy in the public sphere. They pioneered this field of study against a wall of resistance, and the reward for their perseverance is now a tangible reality, against a backdrop of environmental problems with oil production, pressure on the coal industry to reduce emissions, and nuclear energy still politically and socially unviable in many countries. Wind and solar energy are finally emerging as accepted alternatives. The politics of solar energy will be dealt with in a later chapter but, even more than at the time of the oil crisis, the public is now embracing renewable sources as the future energy. Concerning the early problems Kaneff takes a resigned view:

> … it doesn't matter what the university does or does not do, but in the short term it makes life unpleasant … administrators generally are not up to date – they don't have the vision, they don't appreciate the problems. Who can blame them? (Kaneff, interview 2008)

3. The Big Dish: Looking beyond the 1990s

The Big Dish, the solar concentrator section of the solar thermal power system officially designated SG3 (Solar Generator 3), sits on the bank of Sullivans Creek at ANU, below the hill that has been occupied since the 1950s by the buildings of RSPhysS. The Big Dish was constructed in the early 1990s. It has been joined by a second, slightly larger, 'big dish' that was funded as part of the ANU–Wizard relationship in the late 2000s. The dishes are visible to cyclists and lunchtime walkers on the Sullivans Creek bike track, to rowers on the creek itself, and in brief glimpses to winter traffic on Parkes Way, when screening trees have lost their leaves. They attract attention only when used as a backdrop for the occasional media event, such as the 2007 launch of the ACT Government's climate change strategy. The dishes' mirrored surfaces, focal point receivers and hydraulic tracking systems combine to represent the culmination of hardware evolution over two decades of research in solar energy. The SG3 dish, the largest paraboloidal dish solar concentrator in the world, was conceived, designed and configured by Stephen Kaneff and his team of researchers and technicians in the Department of Engineering Physics, RSPhysS over the period 1986–94, as a prototype for large-scale solar power generation.

Funded by the NSW Government and Newcastle-based Allco Steel, and negotiated directly between the department and the NSW Department of Minerals and Energy with funds channelled via ANUTECH, the SG3 project was, like White Cliffs, subject to various political machinations. In 1986, Premier Neville Wran resigned and Barrie Unsworth was appointed premier, with a state election to be held within two years. The then NSW minister for mineral resources and energy, Peter Cox, approved funding for the SG3 project in late 1986, after a visit to the White Cliffs solar power station. Cox had a good grasp of the technology involved in the solar collectors and energy conversion system, and he had the vision to see the long-term potential of large-scale solar thermal projects, but there were inevitable bureaucratic delays in the approval process. The state election was held in March 1988, resulting in a change of government. Neil Pickard, the new minister for mineral resources and energy in the incoming Liberal government led by Nick Greiner, continued the funding process, which was finally approved and implemented by the Greiner government in December 1988. Kaneff recalls that it took from 1986 to 1990 to get work started on the Big Dish, albeit with a reduced budget and with the major hold-up having been the funding hiatus. It was gratifying that the new conservative government was willing to fund continuation of the work.

Design and development

The Big Dish was conceived in early 1986 out of a need to overcome some of the technical difficulties encountered in a small community solar power project for the island of Molokai, in the Hawai'i group of islands, which the Department of Engineering Physics carried out over the period 1984 to 1988, in collaboration with the US firm Power Kinetics of Troy, New York State. This project, funded by the US Department of Energy was set up as a result of engineering physics and Power Kinetics winning a joint tender in response to a statement of opportunity that had been released by the US Government. The tender was won in competition with US companies and universities and its success was a direct consequence of the success of White Cliffs. The Molokai project worked technologically, but was not economical in relation to the 295-square-metre aperture solar collector. It was evident well before the collector had even been built that large collectors could be practicable but that the particular configuration employed for the square aperture Molokai collector would be unsuitable for economical power systems. With this in mind, in early 1986 Kaneff conceived a new configuration for large collectors which would allow the construction of lightweight structures of great rigidity to ensure that the focal region characteristics would not change significantly, irrespective of the collector orientation or the wind velocity while the solar tracking process was in operation. The new configuration was first built as a scale model to ascertain its physical rigidity and practicality and then checked by Kaneff for a large collector by employing Southwell's 'relaxation method', which had been developed much earlier for calculations associated with the design of aircraft frames, and a desk calculator. This process took some time but revealed that the configuration was viable and could lead to economical collector systems. At a later stage, a computer program, also used in complex frame design, was employed to design the 400 square-metre-aperture collector for the SG3 power system.

While funding was being negotiated for the Big Dish, resources had also become a problem for the thermochemical research being carried out by Peter Carden. It is possible that the high profile and publicly visible nature of the large solar thermal projects aids in attracting funding for such projects as the Big Dish, while the less visible, but no less important, research on thermochemical solar gets less attention. Carden's work at this time was concerned with storing solar energy by using chemical reactions:

> You've got a chemical reaction at the focus — the hot part — and it puts the energy into chemical energy by making a chemical reaction occur and then using heat exchangers — you can have cold fluid coming in, being heated up with a heat exchange to this point where the reaction happens and then the hot stuff comes back again giving its heat back

to the incoming fluid so it ends up coming out cold but it's got trapped energy chemically now, but it's cold, so you don't have to have insulation round your pipes. You can take it away, you can store it, when you want that energy back you can bring it out of storage and reverse the whole process. You go up another heat exchanger into a hot path, reaction occurs, you get more heat out, heat that can go into some sort of engine or something, then the hot products go down the heat exchanger, heat goes through heating up the incoming fluid and back it goes. So it's a beautiful system in theory, because it gives you the storage and it gives you the ability to transfer energy over fairly long distances without any heat loss. (Carden, interview 2008)

Despite Carden and Kaneff having used personal funds to maintain the patents on the thermochemical work over a number of years, by 1986 the patents had lapsed through inadequate maintenance resources and, while the NSW Government found the Big Dish concept appealing, by this time they had other, more pressing problems and were not in a position to support further thermochemical studies. Carden took early retirement at that time, but he notes with satisfaction that his work was carried on by a more recent generation of researchers at ANU.

Once the NSW Government approved and released the funding for the SG3 project, in December 1988, a core team of researchers and technicians was able to begin work on the implementation of Kaneff's design concept. The key personnel for this project, in addition to Kaneff, were Bob Whelan and Ken Inall, who were both involved in the White Cliffs project, along with Geoff Major and Wie Joe. Inall contributed to the power conversion section of project SG3 while Wie Joe contributed to structural design. After the collector had been built, Glen Johnston undertook a photogrammetric study to characterise the optical quality of the collector. Kaneff led the project and management of all aspects was in the hands of the staff of the Department of Engineering Physics, which later became the Energy Research Centre.

By mid-August 1990 the overall system design was completed and Kaneff was able to present this to the NSW Energy Corporation, which subsequently approved it for installation on the ANU site at Sullivans Creek. However, bureaucratic delays meant that the engine room and solar collector were not operational until late 1992 and, even then, the collector was only manually operational. The computer-controlled hydraulic tracking system was not fully operational until 1994.

Growth and success

Officially commissioned in 1994, the Big Dish was pioneering technology. At the time that SG3 was being developed and built, wind energy was beginning to gain in popularity and acceptance. The economic acceptance of renewable energy technology is measured in terms of dollars per watt. The ANU researchers had to prove that SG3 could compete, not only technologically but also economically, within the competitive framework of the energy market. From an engineering perspective, the Big Dish was a landmark because of its size — the world's largest solar dish collector. It was the prototype for a similar dish at Ben-Gurion University in Israel, which was commissioned on the basis of the success of SG3. As it was a prototype, the researchers were able to sort out the technical difficulties of actuation and mirror panels on SG3, with the other potential difficulty, that of the receiver, having already been largely solved in the White Cliffs project.

Two decades after its completion, the Big Dish can be viewed as the forerunner of a new generation of large paraboloidal solar concentrating dishes. It demonstrated that large solar dishes were technically and commercially feasible. Large dishes with Sterling engines and with photovoltaics for solar conversion have since been constructed. The Big Dish was instrumental in promoting the work of the solar thermal group. Over the years, thousands of visitors have been impressed by its size, impact and potential. Amongst those visitors were representatives of Wizard, a company that subsequently obtained funding from the Commonwealth Government to carry forward the work on Big Dish concentrators. This resulted in the construction in the late 2000s of SG4, a 500-square-metre dish located adjacent to the Big Dish, under the leadership of Wizard and Keith Lovegrove. Viewed together, the pair of dishes — which share many obvious characteristics, but differ in design and construction — is an impressive sight.

4. Evolution: Solar energy research as a distinct entity

The Research School of Physical Sciences (RSPhysS) was established on the Acton campus of The Australian National University (ANU) in 1950, following the appointment in 1948 of the Australian nuclear physicist, Mark (later Sir Marcus) Oliphant, who was then living in England, as its foundation director. In the two years prior to its establishment at Acton, the embryonic RSPhysS operated in Birmingham where Oliphant and a small technical staff began work planning for the new school (Ophel and Jenkin 1996). Oliphant's own area of research and expertise ensured that nuclear physics was central to the development of the school which, when it opened in Canberra, had six departments: nuclear physics, theoretical physics, particle physics, radio chemistry, geophysics and astronomy. When Oliphant stepped down as director of RSPhysS and head of the Department of Particle Physics in 1964, a subsequent review of the school resulted in the establishment in 1965 of the Department of Engineering Physics, with Gordon Newstead as its head under a directive to 'centre [the department's] research program on the unique attributes of the homopolar generator as a source of controlled high current, high energy pulses' (Ophel and Jenkin 1996: 39). Following Newstead's retirement in 1970, Stephen Kaneff was appointed to head the department with the focus remaining on the homopolar generator (HPG or the Big Machine).

The Big Machine

The HPG and its centrality to the department require some explanation in the context of the history of solar energy research. In the early 1950s, after seeing work being undertaken in the United States and Union of Soviet Socialist Republics in his field of nuclear physics, Oliphant decided that the nascent RSPhysS should be tasked with building a proton synchrotron. According to the late John Carver, then ANU research fellow and later a distinguished physicist and professor, there was a lot of excitement surrounding the project: 'we were going to do some quick experiments in nuclear physics, and Oliphant and his team were going to build the proton synchrotron with which we would all win Nobel Prizes by discovering the antiproton …' (Foster and Varghese 1996: 254). A synchrotron is a machine for accelerating nuclear particles to very high speeds. The proton synchrotron was, in fact, the second string project adopted after Oliphant's original vision of a cyclo-synchrotron had been abandoned following a massive cost blow out caused by initial under-costing of the project.

The proton synchrotron itself was a highly ambitious project given that the actual infrastructure of the school was still being built, and Oliphant's budget for this project also was considerably short of the mark. Ultimately the proton-synchrotron project was also abandoned due to cost and the fact that overseas research in the area had progressed to the point where RSPhysS was being left behind. The HPG was a central element of the two abandoned projects.

Until the ANU project, work on proton synchrotrons had involved the use of a series of iron magnets. Limits on the power of the magnetic field that could be generated, however, caused limits on the efficacy of the proposed synchrotrons. Oliphant's idea was to create magnetic fields without iron by generating electric currents in copper, theoretically with no limit to the generated field strength. The HPG, while a more modest stand-alone project than the proton synchrotron, was to have been the 'engine room' for its predecessors. The term 'modest', when applied to major research projects, however, is relative. Work on the HPG dragged on through the 1950s and into the 1960s, with grants from the Australian Atomic Energy Commission and the federal government, and increasing pressure from the latter for completion of the project. Finally, in 1962, and after considerable criticism of the project in the public domain, the HPG was 'put through a series of successful tests, during which it delivered nearly 2 million amperes, well over the design limit' (Foster and Varghese 1996: 255).

Shortly after the successful tests, there was an industrial accident in which a technician was blinded when residue in a can of highly combustible sodium and potassium alloy exploded. While work on the HPG continued after the accident, with altered safety conditions, Oliphant withdrew from direct involvement in the project.

The HPG as a working machine was completed in the Department of Engineering Physics under Newstead. Kaneff joined the department in 1966. Over the period 1964–70 several projects utilising the HPG were established and were either powered by it or were intended to be so powered, including high field magnets and solid state physics (Peter Carden), high powered lasers (Hughes and Ken Inall), macroparticle electromagnetic accelerator (Marshall), plasma physics and the ANU tokamak (Liley and Morton). These projects were path-finding ventures, which spawned other departments in RSPhysS, including solid state physics, plasma physics and laser physics.

First spawned was a Department of Solid State Physics, building on the work of Carden in the high field magnet laboratory and experiments in solid state physics, starting in 1970. This provided the opportunity for Carden to change research directions. The tokamak research activity evolved to become the Plasma Physics Centre, which was associated with other Australian universities; the laser research likewise formed and nurtured the Laser Physics Centre, while the macroparticle accelerator work was taken up by the Australian Defence Science Laboratories and then, more substantially by the mid 1980s, in the United States, where it seeded 19 major projects. The self-organising control aspects of Kaneff's work had over the years developed into broader computer based studies and eventually evolved to another full department — the Computer Science Laboratories which, later still, became associated with the new ANU School of Information Sciences.

These evolutionary processes occurred gradually and sequentially over the period 1970–88, and later. At the start of this period the engineering physics constituted one department of seven in RSPhysS. At the end, the total activities generated by the department comprised five departments, representing about one half of the total number of departments in the school (mathematics and astronomy had earlier separated from RSPhysS).

Interestingly, in spite of (in Kaneff's view) substantial beneficial developments of world significance occurring specifically in relation to the HPG generated studies, in the 1976 review of the school, it was reported to the faculty board that the review committee believed: 'continuation of this optimistically contrived arrangement is no longer appropriate nor justified' (RSPhysS 1976: 3.3.2), and work on the HPG ceased. By that time, Kaneff was the head of the department.

Departmental watershed

The solar studies, which were formally initiated in December 1970, allowed Carden, on his return from the United Kingdom and fired with the enthusiasm to pursue research in solar energy, to begin consideration in 1971 of likely profitable paths for the work of the new group in the department — the Energy Conversion Group. Within the first year, Carden had identified solar thermochemical energy conversion as a profitable path to follow. Solar thermochemical energy storage possessed the major advantage that it potentially facilitated the mass utilisation of solar energy. This remained a major objective of the department's solar work.

These developments were a timely catalyst for Kaneff to begin steering his department in the new direction of solar thermal and thermochemical energy, which evolved with thermochemical research under Carden's direction. Simultaneously, other aspects were essential to achieve the mass utilisation of solar energy: solar thermal concentrating collectors and the development of complete power systems, driven mainly by Kaneff, formed a parallel path. Departmental funds allowed researchers Owen Williams (physicist) and Winston Revie (chemical engineer) to be appointed and Bob Whelan, previously the manager of the magnet laboratory, to join the group; followed shortly thereafter by the arrival of a PhD student, Lincoln Paterson, who investigated high pressure storage of gases in underground aquifers as a potential means for storing solar energy on a large scale and long term. Significant government funds were secured to carry on experimental programs during the early years.

The 1976 review of RSPhysS was a catalyst for further change. The review process was the subject of some criticism and it was suggested that, rather than being a model for future university reviews, it 'provided more of a counter-example by highlighting pitfalls to be avoided' (Ophel and Jenkin 1996: 41). The major problem appeared to be that the instigator of the review, Robert Street, was both director of the school and chair of the review committee, and so the likelihood of his being subject to conflicts of interest was high. The circumstances of the review's commission were also unusual: it was commissioned by the faculty board of the school, and therefore the report was 'not accessible to external parties, including the vice-chancellor and Council, without the express approval of the Faculty Board' (Ophel and Jenkin 1996: 41). The committee itself comprised mainly senior academic staff of ANU and some representatives from other Australian universities, but with no fully independent international representation.

It was always Kaneff's understanding that one task of the review was to ensure the cessation of work related to the HPG, and that this was set, almost as a directive to Street, at the highest administrative level. According to Kaneff, the world class results and clear external benefits of the HPG research meant that work could not stop at that time, but years later lack of funds did see the end of this work. The formation of the Plasma Physics Centre and laser and computer science groups was a part of the evolutionary process of growth that saw the era of HPG research come to an end. Well before the Plasma Physics Centre developed, the Department of Solid State Physics was created to take advantage of Carden's high field magnet work. At the time of Street's review, however, the only program terminated was that associated with the macroparticle accelerator work. This could not proceed in any case because the next step was to build a much larger experiment for which a new building was needed as well as considerable experimental funds. While it may have been Street's objectives to close down solar energy research in spite of good reviews of the solar work and HPG experimental results, the report marked a watershed for the department.

A later review of the report from within RSPhysS concluded that:

> There is one member of the academic staff assisted by two professional staff engaged in this (solar) work. Some of our assessors have found it difficult … to accept the claims of world priority and uniqueness of concepts. The Committee has most serious doubts that a project of the diversity and complexity described to us is best pursued or is even appropriate in the department. For example, the department's submission described

necessary work in areas such as mirror field control, reflective surfaces, mirror mounts, ammonia circulation systems, heat exchangers, reaction thermodynamics, investigation of catalysts, materials research, storage in underground aquifers and radiation measurements. (RSPhysS, 1976: 16)

Plainly there was hostility to the continuation of solar energy research from some within RSPhysS. Kaneff recalled in 2008 that, in 1976, there were nine staff involved in solar energy comprising four academic, one professional, one PhD and three technical. Notwithstanding the negativity of the review committee, however, he views the report outcome as having been reasonably favourable for solar energy research. In addition, people across the University expressed support for the continuation of solar energy research following the indication that university administration intended to close down the research. This support culminated in a letter signed by many senior academics arguing for continuation of solar research. Together, these and other factors resulted in the research proceeding.

While the hurdles that they had encountered in pursuing applied research in a university environment defined by pure academic research were not entirely removed, Kaneff believed that the report provided them with the imprimatur to continue their applied research, although with the condition that they found their own money. One of the major points in the report in favour of solar energy research, according to Kaneff, was that they 'were looking at things from an economic and industrial point of view, and not just from a "pure" research point of view'. Nonetheless, Kaneff also speculates that some people may have believed that there would be insufficient external interest in solar energy, and therefore insufficient funds available to make the solar energy group viable, and that as a result solar energy research at ANU would disappear naturally.

Over the period 1980–88 the funding for the Department of Engineering Physics acquired from outside sources increased greatly and the research became increasingly solar oriented. In 1987 the name of the department was changed to the Energy Research Centre (ERC) within the Research School of Physical Sciences and Engineering (RSPhysSE, the name of the school itself also having been changed in accordance with activities accepted as inclusive of engineering).

Solar thermal and thermochemical energy studies in that period were still not favoured by the University or the school. The obligatory review that was traditionally held prior to retirement of a head of department, stated that 'it would be difficult to find a new Professorial Head of the Energy Research Centre and its programs' (Kaneff, personal records) on the retirement of Kaneff, and recommended closure of the operations of the department at the end of 1991. The strong support of eminent solar researchers in the United States and Europe, which was presented to the 1991 review, urging ANU to continue the solar work, caused the University Council to determine that the work of the centre should continue as part of ANUTECH. Kaneff, as emeritus professor, continued as head, to enable the solar and related technologies to develop further and be commercialised.

In the event, after 1991 RSPhysSE still provided laboratory space and some resources to the centre until the end of 1994, as well as extended visiting fellow status within the school to the existing scientific staff of the ERC. But the ERC was, by ANUTECH decision, to be a 'Profit Centre' within ANUTECH; and research and development would proceed only where external funding could be sourced. Because ANUTECH did not wish to be involved in thermochemical or phase change storage studies, these were handed over to the new Department of Engineering which, from 1993, formed its own solar thermal group and continued to use the original laboratory space and ERC facilities in RSPhysSE; a situation that continues today.

ANUTECH and the Energy Research Centre

Kaneff notes that, by the time the ERC was transferred administratively from RSPhysSE to ANUTECH, the arrangement being contractually operational from February 1993, the major basic thermodynamic and thermochemical work initiated by Carden, who retired just before the changeover, had been established and studied experimentally in the laboratory. Assessment of the work and programs had already previously produced favourable reports from chemical industries in Australia and Europe, encouraging the continuation of the research. In 1994, a first solar-driven thermochemical loop based on ammonia was devised and demonstrated by PhD student Andreas Luzzi, working with Keith Lovegrove and colleagues in the thermochemical laboratory established by the ERC, thus providing a good introduction for the new Department of Engineering to resume where RSPhysSE and ERC had left off. Over the period 1971–94, major thermochemical contributions, apart from the key contributions of Carden, were made by Whelan whose continuous management, advice and design contributions enabled the research to proceed through many modifications and a major change of laboratory location.

Parallel development of concentrating collectors and of complete solar thermal power systems by Kaneff and his team fared well, resulting from successfully originating large concentrating collectors and the concurrent evolutionary installation of solar thermal power systems at White Cliffs, Albuquerque, New Mexico, and Solar Generator 3 in Canberra. The relatively short-term nature of these projects encouraged the flow of substantial resources which, by the time the ERC was transferred to ANUTECH, had demonstrated and validated configuration, design and basic viability of large collectors and revealed the potentially attractive economic advantages of the technology in practice; and its application in complete solar thermal solar power generation systems, as well as producing successful approaches to solar power system realisation. This allowed ANUTECH to start exploring commercialisation of the technology.

Funds were forthcoming from various sources, particularly the Northern Territory Power and Water Authority, which also commissioned Kaneff, via ANUTECH, to carry out a prefeasibility study for a four-megawatt addition to the Tennant Creek diesel power station. The requirement was to replace some of the existing diesel generators by two megawatts of solar thermal electricity, based on large dishes and two megawatts of gas turbine generation. The study, completed in 1992, was favourable to the proposal and attracted other Australian power utilities to join a consortium, together with the National Energy Research, Development and Demonstration Corporation (NERDDC), to take the proposal further. As a result, a feasibility study was carried out by ERC in 1993–94 in cooperation with all parties, and the report indicated that the project was viable.

In June 1994 John Hansen, chief engineer of the Plataforma Solar de Almería research and development and test facility, provided a favourable assessment of the ERC large collector technology and the Tennant Creek power station proposal, as a result of a commission by the Joint Utilities Consortium. As a consequence, ANUTECH embarked on what turned out to be very protracted negotiations for the sale of the solar thermal technology. Unfortunately acceptable sale terms were not agreed and the joint utilities and NERDDC, in the event, left the arrangement indefinitely inconclusive and did not even agree on the actual power station site; nor were funds for the project forthcoming.

Over the period 1993 to 2005, the ERC continued with the development and promulgation of the technology, effectively working on the programs carried over from the RSPhysSE, with financial support from external sources. This involved developing SG3, the Big Dish solar dish collector and power generation system. Continuation and development of more-advanced and larger solar thermal power systems, and the development of more cost-effective commercially oriented collector systems, were central issues which occupied much of the effort, along with the time consumed in writing and submitting patents to ensure protection for the technology that was being developed. This arrangement provided ANUTECH and ANU with the intellectual property at virtually no expense to themselves, enabling the technology to be readily sold and licensed.

The role of the ERC also involved participating in fund raising and demonstrating and presenting the technology to interested parties. A promising association was formed in 1997 with the construction company Transfield, whereby the

ERC designed a 1-megawatt electrical solar collector array using large collectors to provide steam for a steam turbine for a Transfield plant in Queensland. This led to a proposal with Transfield and Pacific Power for a 2.5-megawatt electrical solar contribution to a large turbine at the Eraring Power Station, which was the basis for a showcase grant from the Australian Greenhouse Office in 1998, although the project ultimately did not proceed.

An important potential application area of the technology arose in relation to combined systems for electricity generation and provision of fresh water. At the request of the Whyalla Council, Kaneff carried out a detailed prefeasibility study in 1998–99 for providing 24 megawatts of solar thermal electrical power together with 20 megalitres per day of desalinated sea water, with the system requiring 200 collectors each of 400-square-metres aperture. The project was supported by the South Australian Government and KPMG were commissioned to carry out a detailed assessment of the proposal.

In 1998–2000 ANUTECH designed, constructed, installed and commissioned a second 400-square-metre Big Dish in Israel: in this case a high optical performance collector similar to the SG3 unit but with a more concentrated focal region to permit a wide range of experimental studies using concentrated solar energy, including solar powered lasers and concentrated photovoltaics (PV). The ERC also carried out several studies for multi-megawatt power systems as well as village power supplies in India during 1994–2002, desalination studies for the NSW Government in 1999 and utilising solar-driven technology to achieve land reclamation in the Murray River Basin at Kerang for the Department of Agriculture and other organisations in Victoria in 2001.

ANUTECH underwent structural changes after the turn of the century, being split into two sections — ANU Enterprise, which was to handle University innovation, and ANU College, which had a teaching function. The ERC was included in ANU Enterprise and functioned until 2005. In 2005, ANU administration made an agreement with the solar energy technology developer Wizard Power, a spin out company from a Canberra computer software company, to advance the technology commercially.

Kaneff stresses that the origination of the solar thermal technology — large collectors and complete solar thermal power systems employing such collectors — was developed within RSPhysS over the period 1970 to 1994 by the Department of Engineering Physics, which was subsequently renamed the Energy Research Centre. For the entire tenure of ERC as part of ANUTECH, the ERC functioned full time in further developing the technology and providing support to ANUTECH to market the technology, in the process considerably enhancing the technology, writing patents and providing information, writing reports, initiating new designs and developments in relation to both collectors and power systems. Frequent presentations were made to industry and power utilities.

Kaneff reports that the functions of the ERC were carried out without any remuneration to ERC members and at negligible cost to ANU, to which the patent rights to the technology were assigned. Not only was a new technology developed, but the rights to the technology were gained almost for free. The ERC was a storehouse of solar energy knowledge and understanding, as well as being aware of the developing potential of the technology, practical improvements possible, and the further development directions needed to enhance economic aspects.

Kaneff feels bitter about later events and argues that it would be reasonable to expect ANU to have called upon his expertise when it came to presenting the technology to interested parties. Instead, the University chose to withhold notice of the negotiations with Wizard Power. The news of the success of the negotiations with Wizard to commercialise large-dish technology reached Kaneff only some weeks after the event and, in addition, it subsequently became known that the board of ANU Enterprises was to declare the ERC redundant. Kaneff found this situation untenable and resigned from his 13-year, unpaid service as head of ERC (ANUTECH) in September 2005.

In spite of difficulties that were experienced during the formative stages of the centre's association with ANUTECH, Kaneff stresses that during the years 1979 to 1996, John Morphett proved to be a successful and respected managing director. He oversaw the required functions in developing and marketing the solar technology, in spite of many hurdles both internal and external.

ANUTECH and the Department of Engineering

A research group for PV was established by Andrew Blakers in 1991 within the new Department of Engineering. Kaneff's retirement as professor of engineering physics in 1991 made him ineligible to be main supervisor to PhD scholars. Two well-qualified scholars wishing to pursue solar concentrator research came to ANU in 1993. Blakers agreed to supervise one (Glen Johnston) and Keith Lovegrove the other (Andreas Luzzi). This provided the academic base for ERC to provide access to laboratories and other facilities to PhD students. Subsequently, Lovegrove, Luzzi and Johnston formed a solar thermal group within the Department of Engineering.

After the formal launch of the Big Dish, the work of Carden on the use of ammonia as a thermochemical storage medium was carried forward by Luzzi during his PhD studies. Ammonia thermochemical storage relies on the chemical properties of ammonia to store heat. Concentrated sunlight heats a reactor filled with a catalyst and hot ammonia. The ammonia absorbs heat and is split into hydrogen and nitrogen, thus storing chemical energy. To recover the stored chemical energy, the hydrogen and nitrogen are placed in a reactor where they form ammonia and give up heat. Ammonia synthesis from hydrogen and nitrogen for fertilisers is a large industry, with little need for additional research, and so the main focus of the work was on the economical splitting of ammonia at the focus of a paraboloidal dish. Two advantages of the ammonia thermochemical energy storage system are that it has no side-reactions and that it is a closed-loop. The unreacted liquid ammonia and the reacted gaseous hydrogen and nitrogen components can be stored together in a pressure vessel at room temperature. Unlike the hydrogen and oxygen resulting from splitting water, the hydrogen and nitrogen from splitting ammonia cannot explode. During the day, thermochemical energy is collected and stored for use at night when the sun is not shining. As Luzzi explains, whilst this process is industry standard and the properties of ammonia have been used in such ways for over a century, the proof of concept using real-life paraboloidal dish hardware was important to demonstrate its use in solar thermochemistry with a fully functional closed-loop solar system.

At this time ANUTECH transferred its ERC interests in thermochemical energy research to the Department of Engineering, which began to use the solar thermal and other ERC facilities. This informal arrangement began in 1994 and worked satisfactorily for seven years. In 1997 the solar thermal and PV groups within the Department of Engineering merged to become the Centre for Sustainable Energy Systems (CSES), while the ERC within ANUTECH continued its work in developing solar thermal collectors and power systems.

The lack of a formal arrangement between the ERC and Department of Engineering solar thermal group, however, eventually led to problems regarding overall control of the solar thermal work. This raised the question of whether they should there be one group instead of two? Lovegrove felt that he should lead the solar thermal work and, after 1995, advocated the closure of the ERC, leaving the new solar thermal group alone to continue its research. Many discussions were held over several years with the objective of Kaneff joining the Department of Engineering solar thermal group as a visiting fellow. Kaneff, however, felt that the conditions proposed would have left him with little opportunity to pursue or influence programs and the negotiations failed. Fortunately, while these discussions were ongoing, joint use of laboratories continued amicably.

At the end of the 1990s and up to 2000, tensions arose. At the time, ANUTECH staff, were engaged in active commercial negotiations in respect of dish technology. It was considered that the use by staff from the Department of Engineering solar thermal group of computers and computer programs developed over the years by ERC had the potential to lead to a breach of confidential intellectual property. ANUTECH removed the computers in a pre-dawn 'raid' on the offices of the researchers involved, along with the confidential programs. The event caused unfortunate antagonism and misunderstanding on both sides, although the people involved now regard it with amusement.

Kaneff's position as a pioneer of solar energy research and its key driver over nearly three decades was such that his retirement from RSPhysS in 1991 left no clear line of succession between three researchers: Lovegrove, Johnston and Luzzi. Together these three people led solar thermal research at ANU until the departure of both Johnston and Luzzi from ANU in the early 2000s. Lovegrove carried on the solar thermal group in the Department of Engineering and formed a partnership with Wizard Power. Together, Wizard Power and the group constructed a 500-square-metre solar dish next to the original 400-square-metre dish on the banks of Sullivan's Creek. Following deterioration of the relationship between Wizard Power and ANU during 2010 and 2011, Lovegrove left the University in 2011.

The academic legacy of Kaneff is evident in the solar thermal collectors and infrastructure at ANU and elsewhere and in the work undertaken in the solar thermal group in the Department of Engineering. Acknowledgement of his contribution to solar energy research at ANU, however, has been curiously absent there. Prior to 1994 he and his colleagues published more than 300 papers. In the early 2000s, the website of the solar thermal group listed only one paper, which mentions Kaneff as a co-author, and none of his (more than 180) own papers; none at all by any author on the landmark White Cliffs project; and only 10 of Carden's, all of which were conference papers. The SG3 Big Dish was explained in detail to many visitors but with omission of reference to the researchers who were responsible for its conception, development, construction, validation and use in the first, grid-connected solar thermal power system in Australia (1994). The White Cliffs project, which remains the first great milestone project for solar energy at ANU and in Australia, was awarded the distinction of being an Engineering Heritage Site, but with little recognition within ANU. Kaneff believes that failure to acknowledge the early projects may ultimately reflect negatively on the University.

As a matter of necessity, solar energy researchers at ANU have continued to be entrepreneurial in seeking project partners and funding grants. Blakers, as director of the CSES, is sure that had Kaneff not paved the way for the commercialisation of solar energy within the structure of the University then the role of the current centre and its work would have been much different, and his job more difficult. It remains that much of the commercialisation of research in ANU and the extent to which it is accepted, indeed welcomed by the University itself, is due to the sheer determination of Kaneff and his colleagues to obtain external funding to continue their research and maintain their department at a time when ANU was against the very concept of university engagement in commercialisation of its research.

In 2000 the solar thermal research group, through Luzzi, was engaged in a collaborative solar chemical project with the Swiss company Ammonia Casale, which was funded jointly by the Australian Cooperative Research Centre for Renewable Energy, the Swiss Federal Office of Energy, the German Academic Exchange Service and Neu-Technikum Buchs. The objectives of the project were to:

> demonstrate closed-loop solar operation of a 15-kW thermochemical energy storage system based on ammonia dissociation/re-synthesis; to investigate the performance of key components of the ammonia-based thermochemical energy storage system; and to assess scale-up of solar ammonia dissociation to pre-commercial demonstration using ANU's 'big-dish' technology. (SolarPACES 2000: 89)

This project was one of a number of similar and subsequent thermochemical projects engaged in by the group over the next few years, based on the Big Dish and involving research collaborators from ANU and Switzerland and commercial funding from both governments and various European industry partners. By 2003 the engineering designs to scale-up the earlier closed-loop system to the Big Dish concentrator were completed with the aim of preparing the SG3 collector as a generic research tool. Joint funding from the ACT Government and ANU in 2004 enabled the proof of concept work to be undertaken which successfully tested 'the operation of a trough-driven ammonia dissociation receiver reactor using Ruthenium on Carbon catalyst' (SolarPACES 2004: 80).

As the Energy Research Centre wound down, solar research in the Faculty of Engineering and Computer Science gained a high international profile and grew rapidly. By the year 2000, staff and students numbered 30, and external research funding in the period 1991–2000 was $16 million. By 2012 the number of staff and PhD students had reached 80, and funding won in the period 1991–2012 exceeded $80 million. Several large projects were funded, including the Rockingham and Bruce Hall demonstration trough concentrator systems and SLIVER solar cell technology in partnership with Origin (Boral) Energy. The Australian Solar Institute was established in 2008 with ANU as a core member, and has resulted in the award of more than $25 million in funding to ANU projects.

Transform Solar, a joint venture of Origin Energy and Micron, is undertaking commercialisation of SLIVER technology. SLIVER solar cells are made on very thin single crystalline silicon substrates. SLIVER technology allows large reductions in the consumption of expensive hyper-pure silicon. In addition, just a few 20-centimetre diameter wafers, when processed via the SLIVER technique, yield enough area of solar cell to cover one square metre of solar module. In a 2007 paper, the commercial application of SLIVER technology was described:

[It has been shown] that it is possible to optimize the Sliver cell fabrication process so that high-efficiency cells can be produced using a simplified processing sequence that promotes high consistency and a very high [energy] yield. Simultaneously, efficient module production via the submodule method can be used, using low-cost equipment and standard PV materials only, to reliably and rapidly produce Sliver submodule units which can then be easily handled in a similar manner to conventional solar cells. (Franklin, Blakers, Weber and Everett 2007: 8)

In 2005 Wizard Power entered into a licensing arrangement with ANU for the commercial development of the Big Dish technology and ammonia thermochemical energy storage. This was a major breakthrough for the solar thermal group. The existence at ANU of the 400-square-metre Big Dish was crucial in demonstrating the potential of large paraboloidal concentrators. Wizard Power's interest was the commercialisation of 'big dish' technology and investigation of its applications in a broad range of fields, including power generation, agriculture, desalination, sewerage and water treatment, and urban heating. In particular, Wizard Power played a major part in the development and construction of a new 500–square-metre dish (SG4), built adjacent to the original SG3, on the bank of Sullivans Creek and completed in 2009.

The new Wizard Power Big Dish 'commercial' design represented by the SG4 facility is based on novel space-frame and mirror panel systems that are optimised for the cost-effective deployment of large (tens to hundreds of megawatt capacity) solar fields. These systems are designed for low-cost and rapid mass manufacture, assembly and installation. A custom manufacturing and dish assembly solution, the 'factory-in-the-field', was developed to enable dish frame manufacture and assembly to take place on the site of the power plant, whilst also allowing the adaptation of the solution to different labour and power infrastructure markets around the world without compromising the quality of the construction or increasing the costs involved. In May 2010 the commercial potential of this work was acknowledged by the Commonwealth Government with the announcement of $60 million funding support for the development of the first commercial-scale power plant using 'big dish' technology, the 40-megawatt Whyalla Solar Oasis Stage 1, in South Australia. The $230 million solar oasis will use 300 dishes in its first stage. The consortium is evaluating the expansion of the plant to at least 200 megawatts in future stages. Hopefully this will be the first of many such plants, as Wizard Power is now engaged in pre-feasibility studies in the United States and India for similar and larger plants.

The Whyalla Solar Oasis project is in many ways the culmination of the work undertaken by Kaneff and his team, who originally proposed the project in the late 1990s. With the involvement of Wizard Power, the work of the pioneer researchers, who demonstrated that research into commercial applications need not compromise scientific rigour, has been decisively justified.

The reputation of ANU as a national and international leader in solar energy research is something that has been built up over the past four decades, owing much to the original nine — four research engineering/physicists and five technical staff — who began to carve out a niche for solar energy research in 1970. The legacy of Kaneff is tangible and measurable within CSES, ANU and the international discipline of solar energy research.

While Kaneff and the solar thermal team were working on the Big Dish, research into photovoltaics (PV) was also making ground. After what Luzzi describes as the 'heydays of Australian PV' in the mid 1980s, when Telecom (Telstra) led the world in the uptake of PV technology, there was a sharp decline in interest — and hence reduced support for further work in the area — when the Telecom infrastructure was superseded and dismantled in the early 1990s. The PV research group at ANU was initiated by Blakers in 1991, along with two doctoral students, Klaus Weber and Matthew Stocks. The PV research group remained wholly a university entity rather than falling under the management of ANUTECH. Whereas ANUTECH owned much of the equipment and infrastructure used by the solar thermal group, Blakers had the good fortune to start at the time when RSPhysSE was planning to establish a photolithographic laboratory. Blakers recalls that:

> I knew how to set up semiconductor process laboratories, having established two previously, so I was given sufficient funds to establish a basic semiconductor laboratory and then I was reasonably successful with external grants which allowed me to populate that with PV-tuned equipment and people and so the group grew fairly quickly to about 15 or so people in the mid-1990s through to the late 1990s. And then that grew up to 25 or 30 people when Origin [Boral Energy] came on board in 1998. (Blakers, interview 2008)

From the outset, the PV research group had focused on multiple research themes, on the rationale that if one project was experiencing problems, there would be others to ensure research continuity. Two themes that have been constant areas of research since the early 1990s are research into silicon materials leading ultimately into silicon solar cell fabrication, and the development and application of thin crystalline silicon solar cells to get around the problem of expensive hyper-pure silicon. The aim of the research is to reduce the silicon consumption per kilowatt from around 10 kilograms per kilowatt to below one kilogram per kilowatt range by using thin crystalline silicon solar cells.

Strength and common ground

In 1995 the PV group broadened to include trough concentrators and PV-thermal hybrid systems. The aim was the supply of both solar electricity and solar hot water from the cooling water of the cells, which are subjected to 30 times normal solar concentration at the focus of the trough, thus making it possible to greatly reduce the quantity of silicon needed per kilowatt of output.

The merging of the Department of Engineering's solar activities into the CSES, in 1997, resulted in an increased technical overlap between solar thermal and PV activities. Research in the design and characterisation of mirrors is one area in which PV and thermal research overlapped. During the 1990s, Johnston worked on the photogrammetric study of the exact shape of the Big Dish collector as a PhD student, under the formal and informal supervision of Blakers and Kaneff respectively. Johnston subsequently won a prestigious three-year Australian Research Council (ARC) postdoctoral fellowship. The focus of the fellowship was the development of mirrors for the trough concentrators being developed in a large project under the leadership of Blakers. The optical characterisation tools that Johnston developed during his PhD were integral to the optimisation of the GOML mirror technology.

The election of a conservative Coalition government in 1996, under the leadership of John Howard, and its lack of interest in renewable energy, had a negative impact on solar research activities at ANU. An early action of the new government was the abolition of the Energy Research and Development Corporation (ERDC), which had been a major source of funding for solar research and development at ANU. There were, however, some positive developments for the ERC, nonetheless.

By the time that Kaneff ceased work at ANU and its controlled entities, 35 years had passed since he first introduced solar energy as a research area within RSPhysS. The physical legacy of the research of the solar energy pioneers is the large-scale projects — White Cliffs, the Molokai Albuquerque Project and the Big Dish systems — but, more importantly, in the application and adoption of their work in industry. ANU of 2005 bore very little resemblance to that of 1971, either in appearance or attitude, as far as solar energy research was concerned. As a key winner of both ARC grants and external funding, the CSES was treated as a serious research and teaching centre within the country's leading university, rather than a technical workshop — the latter having been the perception against which Kaneff, Carden, Inall, Williams and their colleagues had fought. Although Blakers acknowledges that he has faced negative comments regarding the commercialisation of their work, it has been nothing like the opposition faced by the early researchers. For Kaneff, though, the work still continues:

> My ultimate objective is to have most of our energy come from the sun. And I believe it can be done, and I'm not obsessed with it, I know that it's going to take a long time. (Kaneff, interview 2008)

5. The wider Australian solar scene: 1970s–1990s

While Stephen Kaneff and Peter Carden were building their research program at ANU, two other Australian universities were starting to take a research interest in solar energy. In 1974, Martin Green, now professor and executive research director at the Photovoltaics Centre of Excellence at the University of New South Wales (UNSW), returned to Australia — after having completed his PhD in solar energy at McMaster University in Canada — to work with Lou Davies at UNSW in solar energy. He quickly became aware of the work being undertaken at ANU in thermochemical storage of solar energy and made a trip to Canberra to talk to Kaneff and Carden. At around the same time, David Mills arrived in Australia from Canada to work in the solar program at UNSW before moving over to the University of Sydney where another Canadian, Harry Messel, was building on an already formidable reputation in physics by establishing a new solar energy program. The three universities: ANU, UNSW and Sydney, together formed a nexus of solar energy research. Almost four decades later, ANU and UNSW remain leaders in the field.

The work of these researchers, however, built upon Australian foundations going back as far as the 1950s. In the early 1950s, Roger Morse, as head of the Commonwealth Scientific and Industrial Research Organisation's (CSIRO) Division of Mechanical Engineering, developed a simple design for low temperature solar thermal collectors. The design of the final product was described in the CSIRO's magazine, *Ecos*, in 1978:

> A blackened copper panel covered with a glass sheet absorbs the sun's rays and heats up. The heat is then transferred to water flowing through pipes attached to the copper sheet.
>
> Simple… It can't help but heat water, even on somewhat cloudy days. (*Ecos*, 1978: 17)

By the end of the 1950s, in the CSIRO engineering laboratory in the Melbourne suburb of Highett, Morse and his staff had designed a 'flat-plate solar water heater that would be extensively used in government housing in arid and tropical areas' (Baverstock and Gaynor 2010: 5). By the mid 1960s, Morse and his team were focusing on the commercial application of solar hot water systems and, in 1974, he had close to 40 researchers, technicians and engineers working mainly on solar energy research. Morse also undertook the presidency of the Association for Applied Solar Energy (AFASE), later renamed and constituted as the International Solar Energy Society (ISES), from 1969 to 1971.

Meanwhile, the Australian and New Zealand branch of AFASE was establishing itself as the primary vehicle for advancing research and development in solar energy. Part industry association and part scientific forum, the Australian and New Zealand section of ISES (ANZSES) grew slowly but steadily through the 1960s and into the 1970s. The oil crisis of 1973 became the catalyst for the organisation's rapid expansion. In their history of ANZSES (2010), Garry Baverstock and Andrea Gaynor note that between 1974 and 1975 membership of the society grew from 260 to 420, increasing to 764 by the end of 1978. Total ISES membership reached around 8000 by the same time. The oil crisis and its role in the advancement of solar energy research is a recurring theme, with many of the researchers involved in the early days of solar energy in Australia. Baverstock and Gaynor note, however, that during the 1970s much of the industry focus in solar energy was on solar heating of water, a point also made by Mills, who has noted that the company Solahart was one of the first big contributors to solar energy research in Australia. In the area of photovoltaics (PV), Green points out that Telecom Australia was a pioneer in 1978, by adapting solar PV panels from their early applications on spacecraft for use in remote area telecommunications.

Funding and commercialisation

Funding for solar energy research in the universities, however, was inconsistent. Messel's Energy Research Centre at Sydney, established in 1973 to investigate 'the biological side of solar energy by the Departments of Biology, Biochemistry and Chemical Engineering' (Millar 1987: 97), was funded by an internal university grant. Mills recalls that during his time at UNSW, the university was offered a large grant for solar energy research by investors from Arab nations. It was, he said, 'too large' and the university turned it down. Messel at Sydney, however, had no qualms about accepting such a large grant. At the beginning of 1977 he had been informed by the university's deputy vice-chancellor that university funding for the solar energy group was to be terminated that year. Then premier, Neville Wran contacted Messel and offered a NSW Government grant of $1million and assistance in furthering negotiations with Prince Nawaf bin Abdul Aziz al Saud, who provided $5 million to further solar energy research and its commercialisation. This was the largest single grant ever received by the university at that time (Millar 1987: 100) and in a 2011 interview, UNSW's Martin Green jokingly commented that they're still spending it. Messel himself, in a 2009 interview for the ABC TV program *Talking Heads*, commented that he was often referred to as an entrepreneur but that, at the time, 'the word entrepreneur wasn't something that praised you, it meant that you were pretty low scum …' (ABC 2009). Messel, like Kaneff at ANU, was finding that the commercialisation of university scientific research was not universally accepted in the academic domain and indeed, the two spheres of industry and research were regarded, by some at least, to be mutually exclusive.

The research-industry dichotomy was not an issue within ANZSES, except insofar as the organisation worked towards resolving the apparent impasse that existed in the university environments regarding the commercialisation of solar research. Mills recalls that, during the 1970s, everyone involved in solar energy associated through ANZSES, which had a very technical orientation. Monica Oliphant, adjunct associate professor at the University of South Australia and former (2008–09) president of the ISES, points out that initially ISES, and its Australasian section, was a strong, science-based forum. In recent years, however, it has been moving towards becoming an industry lobby group, resulting in some conflict within the organisation. On the one hand, this is a positive development as it indicates that the solar industry has grown to the point where it can legitimately have a lobby group to represent its interests at government level. On the other hand, it leaves solar energy researchers without a unified national and international platform beyond the purely academic sphere. Industry associations tend to focus their energies on such matters as marketing strategies and political alliances rather than sharing ideas about new research.

Expanding the field

During the early days of research, the common ground of ANZSES provided an ideal forum for collegial overlap and exchanges of ideas but, beyond that association, the three universities engaging in solar energy research pursued different paths: the ANU engineering physics group was working on solar thermal concentrator dish collectors and thermochemical research; UNSW worked on PV; and Sydney worked mostly on evacuated tube collectors and selective absorbing surfaces for solar thermal concentrator trough collectors. The work of all three was actively supported by and within ANZSES, particularly its NSW branch. Baverstock and Gaynor (2010) note that the Solar Energy Centre established by the society at The Rocks in Sydney in 1979 attracted almost 2000 visitors in its first week of operation. Under the direction of Roger Gammon, this facility was an information and demonstration centre that brought together the practical applications and latest innovations being undertaken at all three of the universities involved in research. Gammon recalls that in one such demonstration project:

> We put up a whole array of solar collectors on the roof of the Argyle Centre where our demonstration centre was, and we piped that into

a solar absorption refrigerator so we could demonstrate how you can actually cool things using solar energy as well as heat things up. (Gammon, interview 2008)

In addition to providing a public face for solar energy research, Gammon was also responsible for coordinating state-based research funding. Located in the heart of Sydney and coordinating research grants, Gammon was in a position to observe and implement some of the most exciting solar energy projects being undertaken at the time.

While the Solar Energy Centre was showcasing the NSW solar activities, in 1978 the Commonwealth established a body to coordinate projects and funding for energy research at a national level — the National Energy Research Development and Demonstration Council (NERDDC). Baverstock and Gaynor note that renewable energy technologies received a significant boost in 1979, once again with the onset of a second oil crisis 'when prices doubled as a result of events surrounding the Iranian revolution and subsequent Iran–Iraq war, and more price rises were predicted' (2010: 14).

Oliphant was a member of NERDDC, and recalls seeing the evolving solar energy rivalry between the three universities as project proposals were received during the 1980s. The main proponents, in hindsight, do not regard their work during this time as representing rivalry in any competitive sense of the word. Both Green and Mills remember the collegial but competitive relationship between the three universities and their respective areas of solar energy research. The competitive aspect was triangular rivalry based on who could produce solar energy the most efficiently and cheaply. Such a rivalry in the context of producing energy cost-efficiently makes perfect sense: funding under NERDDC was highly competitive and the role of the body was to make recommendations to the federal minister for national development. The potential for a project to generate efficient and cheap energy from renewable sources was crucial to the outcome of NERDDC's decisions. More than the competition for limited funding, however, was the belief of each of the key research leaders in the future of their own research and their own area of solar technology. This belief has been borne out in each case over the ensuing decades.

During Kaneff's leadership of the solar energy research in the Department of Engineering Physics and the Energy Research Centre (ERC) from 1970 to 2005, federal government support (via NERDDC) was of little consequence, except during the very early years when initial support was helpful in progressing the thermochemical studies of Carden and colleagues. However, NERDDC policy changed to exclude long-term research from their ambit. No early funds were available to support the collector and power system studies, but Kaneff was able to secure continuing substantial funds from the NSW Government, together with other funds from industrial and overseas sources, which facilitated the progress of this work.

By 1988 the federal government was recognising the potential for solar energy to 'make a growing contribution in specific market niches' (Department of Primary Industries and Energy 1988: 12.1), specifying solar water heating systems, solar industrial process heat systems and electricity production from solar thermal or PV processes as 'new technologies' that had emerged or shown promise over the previous two decades. Projects such as White Cliffs, which had attracted considerable public and political interest, almost certainly contributed directly to that but, according to Mills, although these were ultimately successful and contributed to the development and promotion of the solar industry, projects such as Kaneff's early flagship were a lot more difficult to get going in Australia than they would have been in other countries. A lot of Australian research, he notes, was done 'on the cheap'. This view was shared by the University's Bob Whelan, who recalled scavenging for materials and recycling parts of previous projects to engineer new ones; and Glen Johnston who described his first research trip overseas and how impressed he was by the state–of-the-art facilities in the German solar labs.

During the 1980s the potential of solar energy became recognised on a far wider scale than within the physics and engineering departments of the universities. In 1981 the Energy Authority of New South Wales published a report on the potential of solar ponds for electricity generation in remote locations. The report's authors recommended 'a demonstration solar pond in NSW with heat for either industrial applications or space heating' (Gerofi and Fenton, 1981), noting further that in 'locations suitable for solar ponds, they can provide power more economically than any other solar driven system' (Gerofi and Fenton 1981: 31).

Meanwhile the UNSW Faculty of Architecture was taking an interest in the potential of solar energy as a passive design principle in domestic applications. The Bonnyrigg Solar Village project began in 1981 with the construction of 12 energy efficient houses in the south-west Sydney local government area of Fairfield, with a further three conventional houses as a control group. All 15 houses were built according to existing NSW Housing Commission planning standards with the intention of the group being added to the commission's housing range at the completion of the project, although the 12 energy efficient houses were of new or modified designs in accordance with the researchers' aims of increasing the profile, knowledge and awareness of passive solar energy in housing (Ballinger 1985: 2). While using only conventional energy sources for the solar village, the design features of the experimental homes on the site resulted in far greater energy efficiency than the control group of three standard commission-design homes (Ballinger 1985: 9). Thermal storage in architectural features such as masonry walls, north-facing windows and clerestory windows, and thermal insulation were found to contribute to overall energy efficiency, features which are now taken for granted in new, five-star energy-rated houses that are constructed today. Contemporary 'green' architecture and design researchers such as Janis Birkeland (2002, 2008b) now routinely incorporate such features in combination with PV and solar hot water systems for domestic, commercial and industrial applications.

The tandem development of solar energy research and its architectural applications, first in solar passive design and then incorporating solar energy systems, was one of the important benefits of ANZSES and the Solar Energy Centre. The forum afforded research scientists and architects opportunities to collaborate in such a way as to ascertain the need for domestic-scale solar energy systems, and enabled the public to view the applications of solar energy. This combination led, quite deliberately, to more successful commercialisation within the sector and an increased political profile for solar energy as an alternative or adjunct to conventional fossil fuel, nuclear or hydro power.

Politically, the early 1980s was a dynamic time for solar energy in New South Wales, which was leading the rest of the country in the research, development and commercialisation of the field. In 1981, Gammon presented a paper to the ISES (ANZ section) conference held at Macquarie University, in which he detailed the NSW Government's program for the advancement of solar energy through the NSW Energy Authority. He noted in his introduction that:

> ... a new technology passes through a number of phases from the time of its initial conception to its widespread utilization. In the case of solar energy, these can be categorized as (i) systems analysis, (ii) research and development, (iii) demonstration, (iv) standards and testing, (v) commercialisation, and (vi) public education and information. The various solar technologies do not necessarily progress in an orderly fashion from (i) through (vi), and State Government assistance with this progression is not necessarily appropriate in each category. For example, research and development programmes are generally expensive and best left to the Federal Government, unless there are particular areas of interest to the State in which it possesses the necessary expertise. Systems analysis, though not highly visible, is often the most cost-effective since errors in policy can be corrected at an earlier stage at less cost. (Gammon 1981: 2)

At the time that Gammon presented his paper, the NSW Government was indeed involved in the process of solar energy research and development. Through the financial backing and assistance of Wran, the ANU White Cliffs solar thermal project was underway as the prototype solar energy power station. Also with Wran's assistance, Sydney's solar energy program was making progress in evacuated tube collectors and was working on transferring the technology to industrial applications. Through the Solar Energy Centre at The Rocks, Sydney was demonstrating a home air-conditioning system using evacuated tube collectors. UNSW, via a joint project of the university, the State Energy Authority and the federal Department of National Development and Energy, had established a solar system test facility and was heating a 50-metre indoor swimming pool with '300 square metres of unglazed polypropylene solar collectors' (Gammon 1981: 5), with the State Energy Authority providing additional funding for monitoring. UNSW had also, through a state government grant, developed optical systems capable of maintaining the concentration of sunlight on absorber surfaces without continuous tracking.

Setbacks and politics

The 1990s were less productive for Australian solar energy research. Green believes that having an ARC Centre of Excellence insulated his team to a degree from the backwards slide that occurred in funding and recognition. At the same time, many of their capable Chinese researchers accepted generous offers to return to China to develop solar energy-based businesses there. He notes that around a quarter of the world's PV production is now based in China. The senior officers of many of these companies spent time at UNSW. Andrew Blakers at ANU was just relieved to be able to hold his research group together during that time, and Oliphant also commented that the stop-start policy approach was damaging to the development and commercialisation of solar energy.

Mills believes that it is impossible to separate solar energy research from politics, a view shared by many early and current solar energy researchers. Mills, however, compares the political profile of solar and nuclear physics, pointing out that nuclear energy has more political 'clout' because nuclear physicists have the reputation of being 'pretty bright', as well as the link between nuclear physics and weaponry, whereas the 'solar guys are just working with hot water and there are no solar weapons' (Mills, interview 2011). Certainly nuclear physics has featured prominently in Australian domestic politics with the British atomic testing at Maralinga in South Australia, the French testing in the South Pacific, and Australia's role as a uranium producing and exporting nation. During the 1980s, while solar energy was picking up pace in its research and development stage and making ground in Australia, internationally, nuclear physics was at the centre of politics with US president Ronald Reagan's Strategic Defense Initiative — the so-called Star Wars program — the rapid expansion of nuclear arms capability of many smaller nations and the proliferation of nuclear power plants in Europe and the United States. The 1990s, however, saw the issue of climate change rise in scientific circles as a widely perceived global threat, with a similar potential for global destruction to nuclear weaponry but requiring a different sort of global cooperation to mitigate. At a time when the climate of the planet is under threat and national security is as much about being able to feed the population as it is about defending borders, the 'solar guys' have increased in political standing.

6. Solar energy research and commercial expansion

At The Australian National University (ANU), the White Cliffs project was a catalyst for the establishment of ANUTECH, as the commercial division of the University, and also a catalyst for solar energy research and development in Australia. It was a serendipitous — and highly charged — alignment of academic advancement, technical skill, commercial application and political imperative. One of the obstacles facing Stephen Kaneff and his team was that the University still regarded applied science and engineering research as falling outside the parameters of its mission under the Commonwealth Act[1] by which it was established. Applied or technical work was viewed as more in the realm of the technical institutions; thus solar energy research, which had broad application, was accorded a low status on the traditional academic scale.

The reason for the University's reluctance to engage in applied research lay largely in the ethos guiding its establishment and early development phase. ANU was founded in 1946 as a research institution under federal legislation. More than 60 years later it maintains its focus on research, a factor that has placed the University at the leading edge of Australian tertiary institutions and places it consistently among the best universities in the world.[2] Over the years the University has attracted and produced some of the world's leading thinkers across the breadth of the scientific and humanities research fields, numbering six Nobel laureates among its alumni and past staff.

Funding for research

The funding arrangements for ANU, unlike other Australian universities, were the domain of the Commonwealth. The shift in economic thinking that occurred during the 1980s, however, and the increasing demand for academic 'services' in the commercial sector, resulted in a change in how universities sourced and managed their funding. As the only Australian university created under federal legislation, ANU had always been treated differently in the funding equations from its state counterparts owing to receipt of block funding for the Institute

---

1   The *Australian National University Act 1946: An Act to establish and incorporate a University in the Australian Capital Territory* subsequently repealed and replaced by the *Australian National University Act 1991*.
2   ANU is consistently ranked as the top Australian university in international ranking surveys.

of Advanced Studies. This academic legacy continues today, in that ANU is far stronger in 'pure' than 'applied' research. Summing up the shift in university financing, the first CEO of ANUTECH, John Morphett, wrote in 1990 that:

> Clearly the universities as a whole now have a responsibility for raising a significant percentage of their income from other sources and pay masters. Predating, but now parallel with, this emerging financial pressure has been the recognition that universities have more to contribute to society than just undergraduate and postgraduate teaching and research. (Morphett,1990: 1)

The commercialisation of research alluded to by Morphett, and the view of research and development as a marketable product, quite common and accepted in the contemporary university environment, was largely untried at ANU when Kaneff entered into partnership with the NSW Government for the White Cliffs project.

The strings attached to external funding were not the only reason for the University's attitude towards the commercial application of research carried out under its imprimatur. Morphett described the attitude as 'old university thinking' or more specifically, the maintenance of the integrity of the academic establishment as a separate entity from the murky world of politics and commerce. The ironic reality is that ANU, as an entity created by legislation, had always been subject to political influence, albeit in most cases covert. While an early decision was made to ensure the autonomy of the University, Mark Oliphant himself, as an academic advisor to the University Council in the early 1950s, had cause to complain to a Council member, H.C. 'Nugget' Coombs, about perceived government interference in the appointment of Leslie Melville as vice-chancellor in 1953, over the academic council members' preference for an alternative candidate. The chancellor at the time, Lord Bruce (former Prime Minister Stanley Melbourne Bruce), later admitted to Howard Florey that Prime Minister Menzies had indeed intervened in the appointment (Foster and Varghese 1996: 118).

Oliphant regularly expressed his misgivings about cabinet and treasury interference in the running of the University, describing the University at one point as 'just a very minor government department' (Foster and Varghese 1996: 119). The concept of the University as a beacon of academic independence was at the heart of its academic founders. Upon his appointment in 1948, the first vice chancellor, Sir Douglas Copland, wrote to the prime minister, that 'the establishment and maintenance of academic freedom is more important than the actual research and teaching done inside the walls of a university' (Foster and Varghese 1996: 113). While Foster and Varghese speculate that this statement may have been an exaggeration on Copland's part, Oliphant and his academic colleagues certainly believed that academic freedom was a cornerstone of the institution and that such freedom could only be maintained as long as the University resisted the influence of outside interests.

It was largely as a result of this way of thinking that the gap between the corporate–industrial and the academic worlds remained unbridged for so long. While Foster and Varghese point out that there was vague talk in the 1940s of how industry would take academic 'discoveries' and develop them 'for the benefit of the whole nation' (1996: 356), there was no real idea of how this was to happen, nor any apparent willingness by either party to engage with the other. Academics such as Oliphant remained trenchant in their pursuit of academic independence and were disdainful of the profit motive of industry, while industry dismissed academics as ivory tower dwellers with no place in the world of commerce and industry. The University of New South Wales (UNSW) had a commercial arm in the form of Unisearch, which had been operating since 1959 in accordance with the university's charter to 'aid by research the practical application of science to industry and commerce' (Foster and Varghese 1996: 358). During the 1960s and 1970s, many of the larger institutes of technology followed suit. Wing (1993: 3) notes that the 'more easily exploited areas of consultancy [were] Engineering and Science'. During the 1960s the institutes most likely to engage corporate partners and funding bodies were, however, those described by Wing (1993: 3) as 'institutions not specifically funded for research'. Increasingly, however, this was changing.

The White Cliffs precedent

ANU, with its research-oriented focus, had no precedent in 1978 for handling applied research grants of the size and scope of the White Cliffs proposal and determined that it should maintain a distance from the actual commercial arrangements. At the instigation of Ian Ross, then deputy vice-chancellor of ANU, a company in the model of those already existing at other universities was subsequently established to fulfil that role. As discussed previously, in August 1979, ANUTECH was incorporated, with Morphett, former manager of the RSPhysS laboratories, appointed as its founding manager. Kaneff recalls that it was timely for ANU to have such a facility, as other institutions already had such commercial arrangements, for example UNSW's Unisearch and the Royal Melbourne Institute of Technology's Technisearch. The establishment of ANUTECH followed a slightly different path from its counterpart organisations at other institutions, but it nonetheless provided the corporate and technical structure that the University required to remove itself one step from the commercial arrangement of the White Cliffs project.

While ANU had no precedent for handling large commercial projects, there was, however, already an existing representative structure for commercial divisions within universities. The Australian Tertiary Institutions Commercial Companies Association (ATICCA) was a professional body comprising the 'leaders of those organisations whose purpose is to assist in some way in promoting effective interaction and technology transfer between their host tertiary institutions and commerce, industry and government' (Wing 1993: 1). It was officially founded in 1978 with seven members. The concept of commercial funding for applied research was rapidly gaining ground in the tertiary sector and the arrival of ANUTECH as the University's commercial wing brought ANU into line with a growing number of universities around the country as well as internationally. Taking UNSW's long-established Unisearch as his example, Wing (1993: 2) notes that universities benefited in being able to keep 'commercial activity at arm's length while enjoying its fruits'.

ANUTECH's establishment as the institutional driver gave the White Cliff's project a solid foundation of institutional backing without the University needing to be directly involved. This also, however, threatened to separate the researchers, who were used to having autonomy over their work, from the management of their own project. As noted in Chapter 1, the appointment of Ken Fulton as project manager caused divisions within the project team and, although Fulton's likeable personality and professional competence won over the academic side of the team as the project developed, the outside imposition of a project manager was a deeply unpopular move.

Morphett's technical background as laboratory manager for the research school and his organisational skills as a former army officer proved to be the right combination for bringing ANUTECH and its first commission, the White Cliffs project, to fruition. As late as 1990, however, Morphett considered that the organisation was not fully realising its potential as a 'specialist link' between 'traditional university administration' and commerce, industry and government:

> ANUTECH is the largest such company in Australia and, with the ANU, provides world leadership in its field. Yet ANUTECH is not accepted universally on the campus nor is it anywhere near as useful to academia as it should be. Enlightened and flexible policies on both sides should ensure acceptance that the community's need for commercial academic products need not interfere in any real way with the teaching and research aims of the University, but should complement them. (Morphett 1990)

Since 1979, ANUTECH has been the commercial partner for several more ANU solar energy projects, sometimes controversially.

Controversies

ANUTECH's commercial ownership of the work undertaken as part of projects under its auspice was jealously guarded. Kaneff recalled in 2008 that, during the research and construction of the Big Dish, the University's media services department took no photographs of the projects and that the only photographic record of the projects under construction are the personal records of the researchers and technicians involved. On one occasion, he remembers, the university challenged the use of his own photographs in a publication on the grounds that he was using photos of university property. The university's Media Services Unit was instructed in 1980 to discontinue its photographic documentation of the construction of the hardware components for the White Cliffs project and later, Kaneff says, there were no university photographs taken of the completed White Cliffs solar station or the Big Dish (Kaneff, interview 2008).

In the view of Ray Dicker, who worked with Bob Whelan as a technician on the project in the early 1990s, the perceived reticence on the part of the University to promote the work of the Energy Research Centre (ERC) stemmed from the perception of the project as development rather than research. Dicker recalls:

> We were based right next to the Physics building and the Mathematics building where there was some top-end stuff going on, and here we were scavenging the local tips … to find parts to actually build some of the components on this dish. (Dicker, interview 2009)

These existing divisions deepened upon Kaneff's official retirement in 1991 and the subsequent review of the ERC by ANU that controversially recommended that the ERC be shut down. Fortunately, by shifting the entire ERC staff to work on ANUTECH's existing projects, including SG3, the centre was kept intact with Kaneff being given emeritus status and remaining nominal leader without financial support from the University.

Occasionally, reporters and journalists gained an inaccurate understanding of the development of solar energy research at ANU. An example of this misunderstanding was a 2009 article on the ANU–Wizard Power solar project in to is talkingf coursehe University's campus publication, *The ANU Reporter*. The article states that, '… [Wizard] turned to the Solar Thermal Group in the ANU College of Engineering and Computer Science, proponents and creators of the original big dish' (Couper 2009: 16). The reporter had not understood that this was a misattribution, and that it was the ERC, under the leadership of Kaneff and management of ANUTECH, that had conceived and built the Big Dish. After the report went to print, Kaneff contacted the writer, who amended his report for the online edition of the publication to reflect the accurate story of the Big Dish. It was too late, however, to change the print edition, which had already been widely distributed.

During the 1970s and into the 1980s, the attitude of the University and, indeed, much of the broader scientific community, towards solar research remained lukewarm at best. Whelan (interview 2008) commented that, from his perspective, ANU thought the solar research group were 'hackers and hobbyists':

> We weren't in the same echelon as the high intellectual endeavours that they were pursuing. In a couple of the reviews it's implied that there was no place for this sort of work in the ANU, which was a bit disheartening, because we had people at the forefront of control technology, we were right into catalyst reaction rates and things like that. (Whelan, 2008 interview)

Commercialisation and solar energy expansion

Controversies and doubts aside, in 1996, even before the official launch of the Big Dish, ANUTECH announced the commercial sale of hardware based on the developed technology: a $395,000 contract to construct in Israel a solar thermal dish similar to Kaneff's SG3 prototype. At the time, the Big Dish was described as 'the largest freestanding, steerable, paraboloidal collector in the world' (ANUTECH Update, 1996). In September 1994 the Big Dish was opened to the public as part of National Engineering Week and, shortly thereafter, ANUTECH reported that ANUTECH's ERC in conjunction with the development consortium consisting of the Energy Research and Development Corporation (ERDC), the Northern Territory Power and Water Authority, Pacific Power, the NSW Office of Energy, the Electricity Trust of South Australia, the Queensland Electricity Commission and the State Electricity Commission of Victoria, were planning to construct a 'demonstration solar thermal plant at Tennant Creek, based on multiples of the Big Dish, utilising a steam turbine with a net electrical output of 4MW, with 2MW of solar contribution' (*ANUTECH Update* September 1994). It was noted at the time that ANUTECH was the child of solar technology, formed, as it was, to manage the White Cliffs project and, by 1994, it had become a 'parent to the technology and its commercialisation' (*ANUTECH Update* September 1994).

In 1991 photovoltaic (PV) work at ANU began after the arrival of Andrew Blakers. The work of this group expanded rapidly and, by the mid 1990s, had extended into solar thermal troughs. The formation of the Centre for Sustainable

Energy Systems (CSES) in 1997 provided a firm foundation for subsequent work in the area of solar energy at ANU, and allowed the pioneering work of the ERC in dish concentrators and ammonia thermochemical energy storage to continue.

Despite the sometimes insecure nature of the commercial aspects of solar energy research, the attitude of ANU towards this area of research has improved considerably from the days when technical and applied research were regarded as less than worthy of an academic research institution. Although Blakers has described the university attitude towards solar energy research in the 1990s as 'indifferent', this is still an improvement on the disdain with which it was held in the 1960s and 1970s. He added, however, that while the work being done in solar energy by his department would be held in far higher regard at a lesser university: 'The ANU has many, many good groups and there's only limited capacity by the senior staff to service and congratulate and applaud all the groups ... we recognise that and we're just happy to be in a good university' (Blakers, interview 2008). Blakers also noted that the University has been slower than some others to acknowledge the opportunities created by global concern about climate change, so the work being done in solar energy research has not always had the recognition and focus it may otherwise have had.

7. Environment, climate change and solar energy

The pioneers of solar energy at The Australian National University (ANU) did not regard themselves as tree-hugging greenies with the primary objective of saving the planet. It is true, however, that concern for the environmental impacts associated with the use of fossil fuels, resource extraction and nuclear energy played a large part in their motivation. Influenced significantly by the oil crisis of the early 1970s and the debate over nuclear energy, as well as the imperative to look for alternative viable sources of energy, they were also committed engineering physicists, intent on furthering the science and technology in their chosen field. With the benefit of hindsight and the accumulated knowledge of the global scientific community about climate change, those first ANU researchers who steered the discipline to its current position can be satisfied with the role that they and their work have played in bringing solar energy to the frontline of the Australian climate change armoury.

While climate change is a phenomenon outside the normal realm of solar energy research, in the light of the political debate surrounding the issue, solar energy must be viewed from that angle as well as from the scientific and historical dimensions of this study. This chapter will place solar energy, and the research that made it possible, into that political context from the national perspective.

Realisation and response

The issue of climate change resulting from human industrial and agricultural activity has been widely recognised since the 1950s. It was not until the early 1980s, however, that it became firmly entrenched as an environmental issue. In 1988, scientists at the first international conference on climate change in Toronto concluded by consensus that alteration of the climate was likely and that anthropogenic factors were strongly linked to that likelihood. National governments began to take notice. The 1992 United Nations Conference on Environment and Development (UNCED), the Earth Summit, in Rio de Janeiro, resulted in the establishment of the United Nations Framework Convention on Climate Change (UNFCCC), a non-binding treaty between participating nations with the aim of stabilising greenhouse gas emissions to mitigate climate change. Parties to the UNFCCC have met 14 times since its establishment, with the Kyoto Protocol being developed at the third Conference of the Parties (COP3). The 15th COP was held in Copenhagen at the end of 2009 with the aim of formulating a global agreement for the period after the expiry of the Kyoto Protocol in 2012.

Australia's participation in action and agreements towards emissions reduction has been mixed. The relationship between climate change and greenhouse gas emissions was well recognised by the Australian scientific community by the late 1980s and the government accepted the science shortly thereafter. The commitment of successive governments to renewable energy as a means of cutting emissions has, however, been disappointing to solar energy researchers. Government programs to reduce emissions from the energy sector, such as the Mandatory Renewable Energy Target (MRET) and rebates for homeowners wishing to install domestic solar energy systems have enabled the government to talk the language of renewable energy and meet community expectations that solar energy is being promoted, while still maintaining a business-as-usual approach to fossil fuel extraction, trade and use.

As a response to the first oil crisis in the 1970s, which, apart from being the catalyst to several solar energy research careers also gave rise to considerable community concern about energy, the Australian Government established the National Energy Research, Development and Demonstration Council (NERDDC) in 1977 followed by the National Energy Advisory Committee (NEAC) in 1978. Thanks largely to the work of Roger Morse and his CSIRO team, Australia was, at this time, a world leader in the commercialisation of low temperature, solar thermal technology. By the early 1980s, however, the price of oil had dropped, community concern about oil evaporated and with it, government interest in alternative energy technology.

In 1987 the CSIRO Division of Atmospheric Research held the Greenhouse 87 conference, the papers from which, contributed by over 100 scientists, were published in 1988 as a volume entitled simply *Greenhouse*, putting the issue into an Australian context. The role of renewable energy in addressing greenhouse emissions was addressed, specifically in papers by Ian Lowe and John Coulter, with the former reflecting the views of ANU solar energy researchers in his finding that Australian spending on research and development in renewable energy was 'relatively weak … despite our notable successes in this area' (Lowe 1988: 606). In a table showing the research budgets for renewable energy among OECD countries in 1985, Australia's commitment of $US8 million lagged well behind the recognised leadership of the United States, Japan and the Federal Republic of Germany as well as others such as Canada, the United Kingdom, the Netherlands, Italy and Switzerland. When viewed as a percentage of the total energy research and development budget, Australia's commitment to renewable energy, at 10.4 per cent[1] of the total, appeared also in stark variance with the percentages of Greece (47.7), Spain (25.1), Sweden (24), Denmark (21.3) and New Zealand (20.4). Lowe concluded that:

---

1   A 1988 Commonwealth Government report put the percentage of the 1984–85 total energy research and development budget going to renewable energy at eight per cent (Department of Primary Industries and Energy 1988: 14.1).

The two broad options to reduce the emission of greenhouse gases are renewable energy and nuclear power. Given current social attitudes toward the other by-products of nuclear energy, it would be preferable to increase support for renewable energy technologies. An important element in expanding the use of renewable energy will be the development of a sophisticated economic framework, taking account of depletion and broad environmental effects of energy use. (Lowe 1988: 611)

Lowe's paper also alluded to the opposition towards renewable energy expressed by the established energy industry in Australia, commenting that a survey taken as early as 1984 had shown 'solar hot water to be competitive in most parts of Australia, despite the claims of some electricity authorities' (Lowe 1988: 607). This opposition was to become — and remain — a major obstacle for the researchers and proponents of solar energy in Australia as recognition and acknowledgement of climate change in the 1980s became a catalyst for action in the 1990s and finally a major area of social concern in the 21st century.

In 1992 Mark Diesendorf compiled a report for the Australian Conservation Foundation, detailing the barriers to the uptake of renewable energy. He found that 48 per cent of Australia's emissions of CO_2, the principal component of greenhouse gases, come from electricity generation, with a further 36 per cent from transport (Diesendorf 1992: 1). These two areas, then, were the obvious targets for emissions reduction efforts. By then the White Cliffs project had already demonstrated the viability of off-grid solar power stations and, indeed, the technology had been significantly refined and extended since that station was installed a decade earlier. Concurring with Lowe's earlier finding, Diesendorf concluded that one of the major 'non-technical barriers' to commercial application of solar energy was inadequate funding of research and development. By example he cited the success of the White Cliffs project and the prototype Big Dish, then under construction at ANU under Stephen Kaneff's leadership, and commented that, following Kaneff's retirement from ANU in 1991 and the subsequent decision by the University Council to transfer the Energy Research Centre to ANUTECH, 'only that part of the group's research which can be marketed rapidly is likely to be supported by ANUTECH' (Diesendorf 1992: 8).

Energy generation and emissions

The introduction of new technology is sometimes vigorously resisted where it threatens the vested interests of powerful lobbies. A 1990 report for the NSW Department of Minerals and Energy stated that 93 per cent of the electricity consumed in New South Wales and the Australian Capital Territory came from black coal 'with a small amount of oil used for boiler ignition and some gas for gas turbines' (Energetics Pty Ltd 1990: 4). The 1988 Commonwealth Government *Energy 2000* report put the figure of electricity supply for Australia as a whole from black coal at a more modest 82 per cent (Department of Primary Industries and Energy 1988: 11.2) and 40 per cent of the national primary energy requirement (Department of Primary Industries and Energy 1988: 9.2). In an address to the NSW Clean Coal Summit in 2008, the director of the Australian Coal Association, Peter Freyberg, announced that, in New South Wales, 90 per cent of electricity came from coal. This indicates a three per cent reduction in coal's share of electricity generation in New South Wales over the period 1990–2008, during which time greenhouse gas emissions from the energy sector increased by about 40 per cent (Garnaut 2008).

This underscores the hegemony of the coal industry in Australia, but it is also starkly incongruent with the rhetoric employed by successive governments in support of renewable energy through reports such as that produced by the Australian Greenhouse Office in 2003, *Renewable energy commercialisation in Australia*. This 70-page report promotes government-supported projects 'commercialising innovative new technologies for the generation of energy from renewable sources across a wide range of applications' (Australian Greenhouse Office 2003: vii).

Further strengthening the role of the fossil fuel sector, in 1990 the Energy Research and Development Corporation (ERDC) was established by the Australian Government to replace the NERDDC, with the main difference between the two organisations being that government investment in energy research had to attract industry funding on a dollar-for-dollar basis. The ERDC itself was terminated without replacement in 1997, beginning, for those researchers, scientists and technicians working in the field of renewable energy, a decade of government inaction on greenhouse emissions and climate change and during which time they saw many of the field's best researchers and technology move overseas.

The link between emissions in the energy sector and the broader issue of climate change is well established and has been widely acknowledged since at least 1992, the year of the first UN Earth Summit. Prior to 1992, the issue was acknowledged by environmental groups and much of the scientific community, but had yet to be fully grasped by governments. The debate that existed within the scientific community regarding the causes and symptoms of climate change, was used by governments — notwithstanding the precautionary principle — as an expedient reason not to take vigorous action to reduce emissions. In the late 1980s and early 1990s the deleterious effects of climate change seemed so long-term as to make any action in the short-term appear to be an overreaction to something that may or may not happen at some time in the distant future.

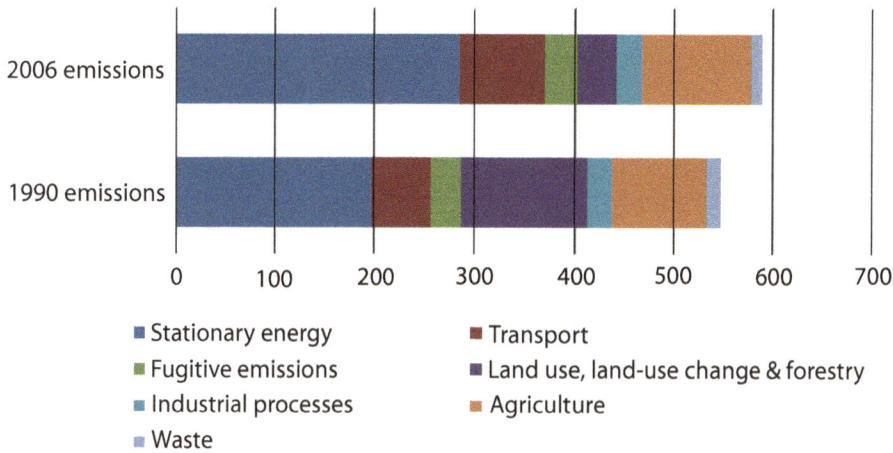

Figure 7.1 Greenhouse gas emissions by sector—2006 and 1990 (measured in Mt CO$_2$-e)

Source: *Garnaut Climate Change Review* 2008.

Earth Summit changed the language and the outlook regarding climate change. On 12 June, 1992, the UNFCCC was signed by 154 countries committed to setting a non-binding aim to reduce greenhouse emissions, with the ultimate objective:

> ... to achieve, in accordance with the relevant provisions of the Convention, stabilization of greenhouse gas concentrations in the atmosphere at a level that would prevent dangerous anthropogenic interference with the climate system. Such a level should be achieved within a time-frame sufficient to allow ecosystems to adapt naturally to climate change, to ensure that food production is not threatened and to enable economic development to proceed in a sustainable manner. (UNFCCC: Article 2 1994)

From the outset, the UNFCCC recognised the role of technology in meeting emissions reduction targets, with Article 4.5 of the convention requiring industrialised and developed countries to ensure that developing countries have access to appropriate technology to assist them to move to a carbon-neutral economy. Moreover, within the diverse and expanding sphere of technology, that related to renewable energy has been central:

> Solar energy and wind-generated electricity — at current levels of efficiency and cost — can replace some fossil-fuel use, and are increasingly being used. Greater employment of such technologies can

increase their efficiencies of scale and lower their costs. The current contribution of such energy-producing methods to world supplies is less than 2 per cent. (UNFCCC 1994)

The case for solar

For Kaneff, the reason for researching and pursuing solar energy was quite clear:

> The reason is [that] you want to get rid of the polluting effects [of coal] … It wasn't enunciated [that way] in the media, but we were not media people. We were in a different world altogether. We [didn't and] still don't want the world to go nuclear. There will probably be a lot more nuclear energy used but imagine what happens if there's another Chernobyl through someone making a mistake, and people do make mistakes. I mean, you make a mistake in a coal-fired power station, well, OK, you have a bit of an accident, but it doesn't last for generations — hundreds of generations. (Kaneff, interview 2008)

In 2011, the catastrophic nuclear incident at Fukushima in Japan, which occurred as a result of an earthquake and tsunami, brought Kaneff's words to stark realisation. As a result of the Fukushima disaster, the Japanese government is now looking to Germany for solar and other renewable energy technology.

The long-term environmental costs of coal and nuclear power, cited here by Kaneff as an incentive for pursuing solar energy, is a recurrent theme in much of the literature on renewable energy. The pollutant effects of coal-fired energy on the environment have been apparent and acknowledged for a long time. The use of coal as the principal fuel burnt in the United Kingdom predates the Industrial Revolution, with coal being used for domestic heat and most industrial processes since the 17th century. The UK *Clean Air Act* of 1956 was enacted as the first piece of legislation ever to deal explicitly with emissions as a means of reducing particulate emissions that had contributed to the chronic problem of dirty air and heavy smog for centuries. The result of the Act was the use of cleaner coals and the removal of power stations to locations outside the urban areas. While this achieved a reduction in particulate urban air pollution, increased visual amenity and improved human health, emissions of carbon dioxide through burning coal for energy were not restricted. These are the invisible emissions which contribute to the greenhouse 'global warming' effect.

Sulphur dioxide, SO_2, is created when the sulphur present in coal is burnt and consequently combined with oxygen during combustion. In the atmosphere it combines with atmospheric moisture to form sulphuric acid — 'acid rain' — which can pollute waterways through acidification. Carbon dioxide, CO_2, is the most prevalent greenhouse gas produced by fossil fuels and, of these, coal releases more than any other during the burning process. CO_2 is colourless and odourless. A vast amount — 28,431,741 tonnes — was emitted into the atmosphere as a result of human activity in 2006, of which almost 76 per cent came from the energy sector (United Nations Development Programme, Millennium Development Goals).

In 1974 *Time* magazine ran a cover story entitled 'Another Ice Age?' which speculated that global cooling was taking place as a result of particulate emissions in the atmosphere blocking the sun's heat from reaching the earth (Eberhart 2007: 221). Coincident with the publication of the *Time* article, scientists were gathering from around the world to launch the Global Atmospheric Research Program in an effort to determine influences on the world's climatic system and the effects, if any, of human agencies. Over a period of years, largely though a study of polar ice cores, they found that temperature variations corresponded with changes in atmospheric CO_2 concentration: as concentration of CO_2 increased, so too did the global temperature.

The case against nuclear

Nuclear power has been promoted for decades by various advocacy bodies and research institutes such as (in the United States) the Nuclear Energy Institute (NEI) and the governments of countries that either have a strong nuclear program or, in the case of Australia, a wealth of uranium available for trade. Nuclear power is billed as the clean alternative to coal, with the NEI, for example, giving a high priority to the promotion of its environmental benefits: 'Nuclear energy is America's largest source of clean-air, carbon-free electricity, producing no gasses or air pollutants' (Nuclear Energy Institute 2009). The problems associated with nuclear energy have been well documented over a number of decades and include the disposal of nuclear waste, power plant decommissioning, the potential use of spent fuel rods in weaponry systems, and the dangers of accidents occurring within nuclear power, stations such as occurred at Three Mile Island (United States) in 1979 and, disastrously, Chernobyl (Union of Soviet Socialist Republics) in 1986 and Fukushima in 2011. As Andreas Luzzi (interview, 2010) adds, commercially very important additional challenges include fuel supply limitations (decreasing ore grades, mine disasters, objections to mining), economic viability without government support (federal loan guarantees, preferential lifelong power purchase mandates, limited insurance liabilities, planning risks), unsafe facilities (fuel reprocessing, fast-breeder technologies), development costs and skills as well as shortages of capabilities (engineering, manufacturing and operation). Kaneff also stresses that what is not usually considered in nuclear energy production assessment is the large amount of fossil fuel burned to gain and process the ores and actually build and ultimately dismantle the power station itself.

The Australian Nuclear Science and Technology Organisation (ANSTO) promotes the use of nuclear energy in Australia by suggesting that Australia will be left behind at a global level. 'Australia stands at the crossroads in deciding its nuclear future. Does it want a backseat or a leadership role?' (ANSTO 2009). The CEO of ANSTO, Adi Paterson, states bluntly on the ANSTO website, 'In my opinion, no modern nation will survive the 21st Century unless it is deeply immersed in nuclear science and technology' (ANSTO 2009).

If the solar energy researchers are in general agreement on one thing, it is that nuclear power is not the answer to the energy question. Kaneff comments that 'we didn't want the world to go nuclear', a sentiment echoed by Peter Carden. Keith Garzoli joked that he referred to 'nuclear energy' as 'unclear energy'. Andrew Blakers noted that, paradoxically, even while there was institutional reluctance to accept solar energy there has been a lot of community support as 'people have been interested in solar ever since nuclear energy became a serious prospect'. Bob Whelan stated that it takes 12–15 years to build a nuclear power station, making it unviable as a source of energy for the immediate future.

It should also be noted that Germany, once one of the world leaders in nuclear energy and the crucible of the anti-nuclear and peace movements of the 1960s, is committed to phasing out its nuclear power stations. Extremely rapid growth in wind and solar utilisation has occurred during the past decade. In recent years, wind and solar power accounted for the largest shares of the additional energy capacity that has been commissioned in Europe and the United States. This trend is likely to strengthen, and spread to other regions of the world. Solar power is now having a substantial impact on the meeting of daytime electricity demand.

Grassroots support

Grassroots and community concern about climate change is a tangible driving force. Over the past few years there has been a proliferation of groups, organisations and networks, all tapping into community concern and anger about what is perceived to be an issue that can be mitigated, if not reversed, through positive action. One of the focal points for many of the grassroots groups is energy use, domestic and industrial, and the promotion of solar energy as the accessible and immediate alternative to fossil fuel is a constant theme. Solar power is perceived to be viable, affordable, safe and easy to obtain as well as maintain. Yet, despite the very popular rebate programs and official words of support for the solar industry, support for rapid growth of the Australian solar energy industry has hitherto been relatively weak, compared with that expressed in other countries. In recent years, however, photovoltaic power system deployment has grown very rapidly, driven initially by attractive feed-in tariffs, and now by the fact that the cost of solar power (without subsidies) has fallen below the retail and commercial electricity tariff in most places in Australia. New wind and photovoltaic installations in Australia are major components of the overall basket of new electric power capacity installed each year.

Whelan (interview 2008) commented that in Israel, 'every house, by law, has to have a solar hot water system. That's it, no mucking around. They look horrible … architecturally they're disastrous but they work.' Queensland University of Technology's Janis Birkeland is a specialist in green buildings and a strong advocate of solar energy in both new buildings and in retrofitting old buildings. She suggests that whole cities can — and should — be retrofitted in such a way that increases their 'net sustainability' (Birkeland 2008: 25). Capturing the energy of the sun through systems integrated into buildings is a major part of reducing the urban environment's dependence on non-renewable resources and overall greenhouse emissions. Many urban planners and local government authorities, however, continue with traditional designs.

Blakers (interview 2008) recalls that he noticed a distinct turning point in Australia for solar energy and the acknowledgement of climate change in 2006, when he was invited to speak at a forum for resource companies:

> … and we had ratbag companies which had all these hired guns and some of the hired guns in the forum listened attentively as I talked about solar energy and I realised that finally we had won the battle to get climate change recognised. That's very different from what has happened in the past — essentially the resource companies, about two or three years ago finally decided that they couldn't deny that climate change was real. Some continue to conduct a rearguard action in a very similar fashion to the cigarette companies but it's no longer overt, but more covert. (Blakers, interview 2008)

It is almost 40 years since Kaneff and colleagues began solar energy research, outside the mainstream research of the Research School of Physical Sciences and Engineering. During that time, scientific understanding of anthropocentric climate change has increased and broadened in parallel with solar energy research. Finally solar energy is being taken seriously as a genuine alternative to the energy status quo and as a means of reducing the carbon emissions contributing to climate change.

8. Solar energy in changing times

A popular aphorism holds that politics is not the business of changing things, but of keeping things the same. In the history of solar energy research, politics has never been far from the heart of the matter and, with it, have been the opposing tensions of progress and stasis — changing things versus keeping things the same. For the purpose of this history, the term 'politics' includes the external political processes, which have served both to aid and obstruct progress in solar energy, and internal university politics that, while a feature of every institution, has been historically instrumental in shaping the direction taken by solar energy research at The Australian National University (ANU).

Progress and process

Solar energy research falls into the category of what economists call 'general-purpose technologies' (Bresnahan and Trajtenberg 1995: 83), which are characterised by their capacity to become widely recognisable, accepted and utilised across society and by their ability to change people's lives (Birrell 2008). General-purpose technologies are 'a single generic technology, recognisable as such over its whole lifetime, [with] much scope for improvement and eventually coming to be widely used, to have many uses, and to have many spillover effects' (Birrell 2008: 2). Among these spillover effects is the capacity of the technology to contribute to the economy through its own production as well as through forming the basis of technological and economic processes beyond itself. Examples of general-purpose technologies that have changed the course of technical and economic history by providing vast opportunities beyond their original technical applications include the internal combustion engine, the electric light, and the internet. Solar energy, too, fits into this category. These things cannot be reinvented (yes, even the wheel was a general-purpose technology!), only refined, improved and applied for different purposes. In the four decades since Stephen Kaneff and Peter Carden established the research discipline at ANU, the use and applications of solar technology have grown exponentially and in tandem with its economic applications. This, in part, forms the basis of the politics of solar energy research.

When NSW Premier Neville Wran provided the funding to ANU for the establishment of a solar power station in a remote town at the western edge of New South Wales, there was the perception of progress: a revolution that would power remote towns by stand-alone solar power stations and bring clean, affordable energy to outback residents generally. Wran was gambling that his vision of progress would be more attractive at the ballot box than his opponents' vision of the status quo. While the Wran Labor government was re-elected in 1981 with an increased majority, it would be simplistic to suggest that the success was substantially due to the issue of solar power. The fact that the government was promoting itself as progressive in embracing the emerging technology and at the same time addressing environmental issues may well, however, have been a factor in the election result.[1] Political tensions between progress and status quo have been a marked feature of solar energy research, providing frustration as well as gain, and defining, in part, many of the directions and outcomes of research.

The political process in a democracy is subject to any number of influences, including non-state actors such as non-government organisations, lobbyists, trans- and supra-national corporations and the economy. As a country rich in mineral resources, Australia has long depended economically upon the strength of the mining industry, particularly in iron ore and coal. The economic paradigm, and its centrality to the political process, is such that, even while Australia is far richer in sunshine than in coal, any potential threat to the coal industry is interpreted as a threat to the economy and, thus, to the whole framework of Australian society. Why is this? Sunshine is an infinite resource and investment in the technology that harnesses its energy as a general-purpose technology for development and export could be a priority of a progressive government.

Economics

The strength of the coal industry lobby rests as much on its history as on its economic muscle. Coal was first discovered in Australia within a few years of European settlement and was being mined in the Hunter district by the beginning of the 19th century. With its justifiable claims of forming part of Australia's industrial, social and economic heritage, and its role as a leading export income earner, the industry has enjoyed a position of political strength in negotiations and decisions about energy. Despite its environmental benefits, and the fact that the original inhabitants of the continent used the sun for energy long before the Europeans started digging up the coal, solar energy currently has a minor economic role. The alignment of solar energy with the environment movement has also served to marginalise it from the industry-dominated political process. The green movement and the Greens political party are frequently perceived and portrayed by their opponents as being opposed to industry, development and progress and confirmed, by inference, as being opposed to the very framework of society. The anti-green slogan 'Greens cost

---

1 In 1980 there was no Greens Party, therefore Greens preferences were not an issue, but green issues were gaining in importance as a concern of the electorate.

jobs' sums up this view. By association, solar energy can be portrayed by its industry opponents as an impractical or idealistic solution to increasing energy needs.

Towards the end of the 1970s, energy costs in New South Wales had risen sharply. This was due largely to increases in fuel costs and interest payments on works in operation (Rosenthal and Russ 1988: 75). At the time when the Wran government was seeking re-election, the issue of energy costs was of considerable political importance. The government needed to take steps to assure the public that the situation was under control — meaning that the government was acting to reduce fuel costs. Solar energy, being independent of the Middle East oil embargoes of the 1970s, was a popular choice for political support. Then, solar energy was not a threat to the coal industry: Figure 8.1 indicates the sources of all energy generated in Australia in 1984–85. By providing funding to ANU to establish a remote solar thermal power station in New South Wales, Wran was playing the political tensions on both sides. He was showing the public that his government was embracing new directions in energy and was committed to providing cheap, accessible electricity to everyone, but at the same time the coal industry knew that solar energy was not going to be a threat to its industry dominance in the near future.

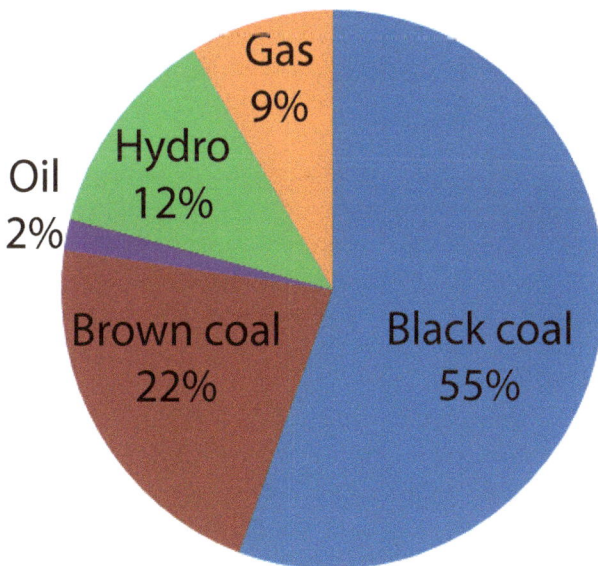

Figure 8.1 Fuel supply for electricity generation, Australia, 1984–85

Source: Rosenthal and Russ 1988.

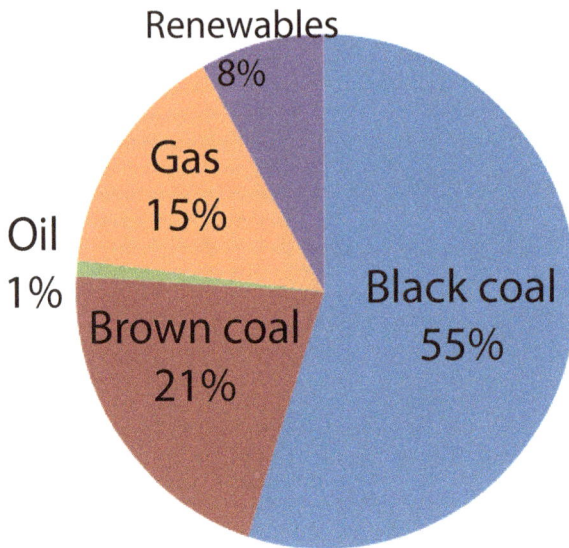

Figure 8.2 Fuel supply for electricity generation, Australia, 2008–09

Source: Australian Bureau of Agriculture and Resource Economics and Sciences (ABARES), Department of Resources, Energy and Tourism, 2011.

According to data from the Australian Bureau of Agriculture and Resource Economics (ABARE), the fuel supply scenario in Australia had not changed much in 2008–09 from the mid 1980s, when Wran was courting both the coal industry and the emerging solar industry. Figure 8.2 indicates just how consistent the coal industry has been in maintaining its position, strongly reflecting the political conservatism regarding energy supply, as well as the fact that, in the absence of a carbon price, coal is a low cost energy option. While the share of the energy market held by the coal industry has remained largely unchanged, the changes within the renewable energy market are relevant to this history. While Rosenthal and Russ (1988: 75) noted that in 1984–85 no significant electrical power in Australia was generated by renewable sources other than hydro, in 2007 the ABARE reported that the 2005–06 renewable sector included power generated from five different significant renewable sources (Figure 8.3), with growth in the sector projected to increase 60 per cent within six years (ABARE 2007: 38). This compared to a projected growth in energy generated by coal of just five per cent, and a total projected growth in the combined non-renewable energy sources of 22 per cent. In 2005–06 non-hydro renewable sources represented less than 20 per cent of the total renewable sources and the ABARE projected figure for 2011–12 predicted a reversal of this situation with non-hydro renewables to make up 83 per cent of the total renewable energy sector. In 2011 ABARES released its figures for 2008–09 showing that, while the non-hydro sources may well fall short of that target by 2011–12, the trend was towards a strong growth in non-hydro renewable, driven by the federal government's target of 20 per cent of electricity from renewables by 2020.

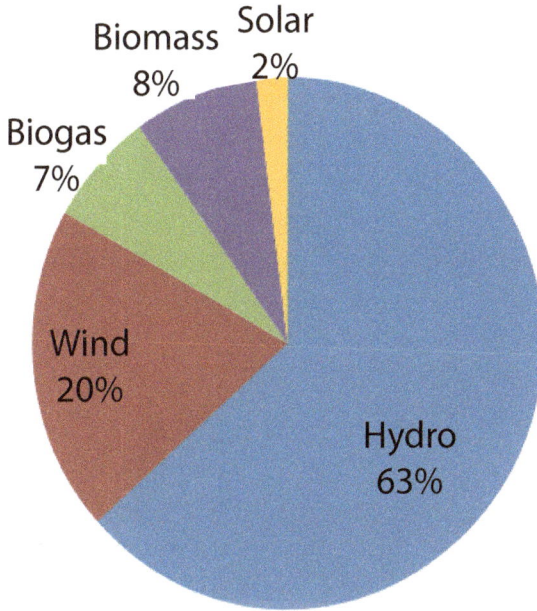

Figure 8.3 Renewable sources of power generation, Australia, 2008–09

Source: ABARES, Department of Resources, Energy and Tourism, 2011.

Timing is everything

In October 1995 the federal minister for primary industries and energy in the Labor government of Paul Keating, Senator Bob Collins, commissioned an Expert Group on Renewable Energy Technologies (EGRET) to research and compile a report on the future and development of the renewable energy sector. Caroline Le Couteur was at the time an assistant director within the Commonwealth Department of Primary Industries and Energy, and became secretary for EGRET. Le Couteur who, at the time of writing, is a serving Greens member of the ACT Legislative Assembly, is the daughter of the late Ken Le Couteur, the founding professor of the Department of Theoretical Physics in the Research School of Physical Sciences and Engineering (RSPhysSE). Of the EGRET report, 'The development and use of renewable energy technologies', she recalls that timing was crucial:

> The election had to be called early in 1996, and with a change in government almost certain, we knew that if the report was not submitted to the Minister before the government went into caretaker mode, it would never see the light of day. (Le Couteur, interview 2010)

The conservative opposition, led by John Howard, had made its position clear on climate change and any measures that might be taken to mitigate emissions and, with noted climate change sceptics within its ranks, it was highly unlikely that the expert group's report was going to be taken seriously, if read at all. Ultimately the EGRET report was submitted to Collins in February 1996, just after the election was called but in time for the minister to release it for public scrutiny.

The group's assessment that the matter would die in the event of a change of government proved to be correct. The incoming Howard government cut off most funding for research and development in renewable energy, refused to accept the Kyoto Protocol at the third United Nations Framework Convention on Climate Change (UNFCCC) Conference of the Parties (COP3), and excluded environmental and pro-solar lobbyists from the decision-making process. Its constant refrain regarding any greenhouse emissions mitigation scheme was that it would not do anything that would harm the Australian economy or cost Australian jobs. It did not elaborate on these claims — elaboration was not politically necessary. By maintaining that any significant measures to reduce emissions would be harmful to the economy and cost jobs, the government effectively maintained that reducing emissions would be economically irresponsible.

The recommendations made by EGRET reflect the major advances made in solar energy research and development over the couple of decades leading up to the change of government in 1996. Of the 15 recommendations (see Appendix 5), not one has been fully realised 16 years after the report set out a program of development for renewable energy which would enable Australia to take a leading position in renewable energy research, development and uptake. In introducing the recommendations, EGRET reported that:

> ... to achieve a sustainable renewable energy industry, an integrated program of market and industry development is necessary. To enable the industry to take advantage of the opportunities that such a program will give it, it is important the Commonwealth Government supports the industry with a consistent and long term program. (EGRET 1996: 38)

Looking at the list of recommendations, Le Couteur comments, 'these were good recommendations — they're still good recommendations and still very relevant. Any of these could still be implemented now — they could still make a difference' (Interview 2010).

The Howard decade

The decade following the election of the Howard government was a difficult time for solar energy research. Andrew Blakers notes that one of the most notable achievements of his centre during this time was merely its survival. While Australian government funding for research in solar energy dwindled and the climate change sceptics on the government benches combined with the coal and business lobbies to ensure that the energy status quo was not only protected, but also encouraged and supported, the position in Europe was reversed. Germany, in particular, forged ahead with solar energy development. Australian research and industry, considered as internationally cutting edge in the mid 1990s, was surpassed by the big research schools and labs in Germany and elsewhere.

Glen Johnston, who undertook his PhD in the 1990s, remembers that his first trip overseas in 1994 made him realise that Australian research in solar energy was highly regarded, but that the huge level of investment in Germany was giving the Germans an advantage in both recognition and results: '... multi mega-millions of dollars, everything's done in stainless steel, workshops full of lathes and milling machines, everything that gets done gets done to the nth degree of precision.' Blakers concurs with this, commenting that the employment generated by the German solar energy industry will eventually surpass that of the car industry. After 1996, the advantage held by European researchers increased because, while the research being undertaken at ANU was still cutting edge and still highly regarded, the lack of political support and government-backed funding for solar research meant that private investors moved towards the areas supported by the government. Johnston's assessment of the situation during this period sums up the frustration felt by all the researchers:

> It's not a big money area. If I'd been researching a cure for cancer, if I'd been researching coal, minerals, mining, perhaps refining different strains of wheat — the big industry players — [there was] money in those areas. But there wasn't [for solar energy]. Solar energy is not a representative area of the industry in Australia. So I'd say that's where the black hole, or the pothole, falls. Because you don't have big industry players behind you. There [are] ones that show interest every now and again, but it's not their core business. (Interview 2008)

Blakers echoes that comment, adding that government resistance to solar energy during the term of the Howard government continued, despite ongoing public support for renewable energy:

> As far as the wider society is concerned, we've always had strong public
> support but that doesn't translate into dollars and it seemed to me the
> stronger the public support for renewables, the more hostility there was
> from the previous government. (Interview 2008)

Recently, however, the rise of large-scale public concern about anthropogenic
climate change has led to a large-scale shift in public and government perceptions
in relation to energy. The defining period in Australia was in the latter half of
2006, when an extended drought, continuously expressed scientific concern, Al
Gore's documentary *An Inconvenient Truth*, amongst other factors, combined to
produce a political situation in which leading politicians could no longer afford
publicly to express scepticism in respect of climate change.

With the election of the Labor government under Kevin Rudd in October 2007,
many restrictions and constraints that had stifled solar energy research since
1996 began to dissolve. During its election campaign the Rudd opposition had
pledged itself to action on climate change and, certainly, its first months in office
indicated a positive start. The ratification of the Kyoto Protocol in December
2007, while largely a symbolic gesture, was a gesture nonetheless that the
previous government had refused to make. The new political climate was reason
for optimism. Indeed, during the 2007 federal election, climate change was one
of four principal themes for the Australian Labor Party. The passage through
the parliament of a carbon tax by the Labor government led by Julia Gillard,
late in 2011, affirmed the shift towards a focus on clean energy. Although the
carbon tax itself was vehemently opposed by the mining and industry sector, a
shift in public perceptions has led directly to an increasing range of projects and
programs in support of solar energy since the 2007 election. The coal industry,
however, retains enormous influence.

Internal politics

As indicated in previous chapters, the internal politics of ANU, and within
RSPhysSE, have served on occasions to hinder rather than help the course
of research. From the University's initial reluctance to accept the work in
solar energy as research when it had a commercial angle, to difficulties with
ANUTECH over project management and ownership, the politics of the research
and development process itself has been as influential in determining the course
of solar research as the external energy politics. By the late 1990s, with Kaneff's
retirement from RSPhysSE and with ANUTECH taking on Kaneff and the Energy
Research Centre (ERC), ANUTECH assumed responsibility for the Big Dish and
its commercialisation. As outlined previously, there was much external interest

in commercialising the technology, but none of the interested parties had the money required to do the job properly and contractual arrangements became mired in red tape and legal problems.

A 2MW Australian Greenhouse Office Showcase Project was granted in 1998, which Andreas Luzzi managed from 1999 onwards as a new, part-time ANUTECH employee. This had the benefit of providing a bridge between the solar thermal group and ANUTECH. The project originally aimed to demonstrate the co-firing of superheated solar steam from an array of 20 paraboloidal SG3 solar dishes into the coal-fired power station at Eraring, in New South Wales. The $7.3 million project was later moved to the Mica Creek power station of remote Mount Isa, in Queensland, where all of the detail engineering for the integration of an array of 18 new 430-square-metre PowerDish™ solar collectors was completed in collaboration with Transfield. The project was eventually cancelled by CS-Energy for commercial reasons.

ANUTECH's involvement in solar thermal research was separate from the new solar thermal group in the Department of Engineering. Keith Lovegrove led the solar thermal group after the departure from ANU of Luzzi and Johnston. The group struggled for funding. Unlike the other solar research groups within the department, the solar thermal group was unable to win Australian Research Council and similar grants, which were the mainstay of solar funding during the Howard years. The thermochemical work of the solar thermal group ended due to lack of funds. The rift between the solar thermal group and ANUTECH had a negative impact on the freedom of operation of the group. Tensions with the rest of the Centre for Sustainable Energy Systems led to the solar thermal group leaving the centre. Eventually the ANU–Wizard Power relationship formed. For two years the solar thermal group then enjoyed rapid growth and a clear focus, until completion of the second big dish at ANU and the souring of the Wizard Power relationship caused an abrupt cessation of funding. In 2011, Lovegrove left ANU, marking the end at ANU of a 'solar thermal group' focused on development of solar dish concentrators.

The early researchers in solar energy were not politically active, but as Bob Whelan says, politics was a necessary adjunct to their survival. 'Steve bore the brunt of tremendous criticism and we felt at times it was because of our perceived political connections.' The EGRET report of 1996 emphasised throughout the importance of the Commonwealth Government taking a leadership role in developing renewable technology, recommending high profile demonstration projects and the creation of a standing committee on renewable energy. Within the University, political support was slow in coming but, once ANU threw its institutional weight behind the issue of climate change and emissions reduction, solar energy research was placed in a more favourable light. In this context, the recent formation at ANU of two sister entities, the Climate Change Institute and the Energy Change Institute, has the potential to forge a strong public profile for low CO_2 emission energy research at ANU.

Blakers gives great credit to Kaneff for playing a major role in changing the way in which the University regards the commercialisation of research. By doing the hard yards, as it were, and obtaining large grants for his work, Kaneff successfully challenged the notion that commercialisation and scientific research were incompatible within ANU. Kaneff and his original small team of researchers, especially Carden, Owen Williams, Ken Inall, and Whelan, through the 1970s and 80s, and their subsequent successors, not only put solar energy research into the public domain, but also enabled successive researchers to seek and obtain commercial applications for high quality scientific research. Their work instigated the formation of ANUTECH and its successor organisations, and changed the research culture of ANU. The current high international standing of ANU in solar energy research, and the widespread social acceptance of solar energy as the energy source of the future, owes much to the determination of a small group of people who followed the sun against the advice of their peers, the research trends of the time and the established system.

Appendix 1. List of people interviewed

Andrew Blakers
Peter Carden
Ray Dicker
Roger Gammon
Keith Garzoli
Martin Green
Glen Johnston
Stephen Kaneff
Caroline Le Couteur
Andreas Luzzi
David Mills
John Morphett
Monica Oliphant
Robert Whelan

Appendix 2. Organisational flow chart

Research School of Physical Sciences and Engineering, 1950–2000 (adapted and simplified from Ophel and Jenkin 1996)

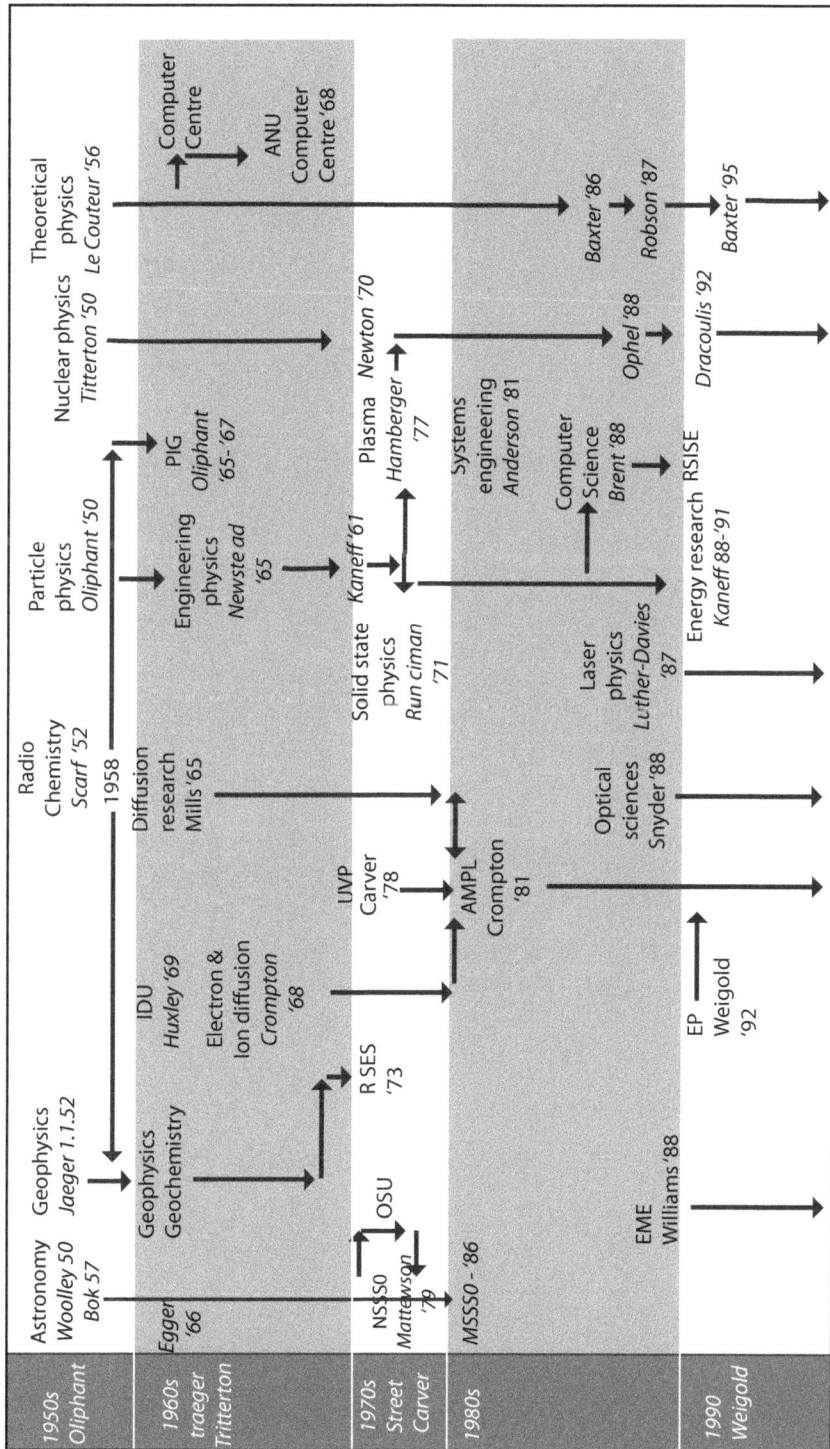

Appendix 3. One page summary from 1970s to the present

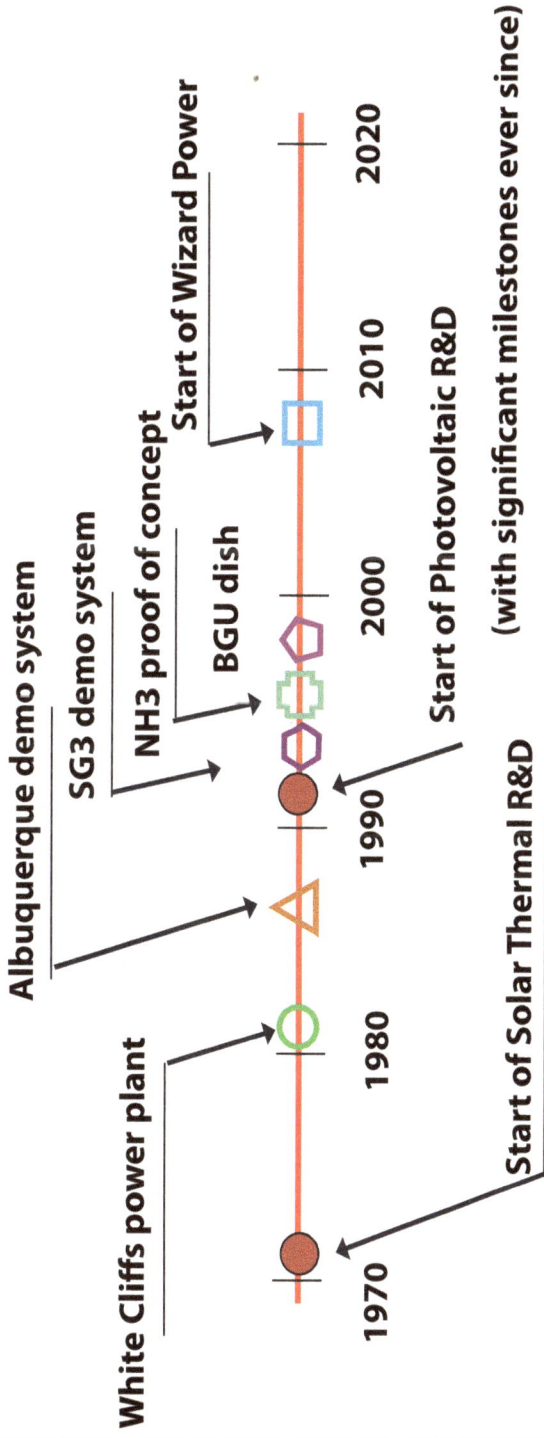

Timeline of historical
Solar thermal **Milestones** at ANU

White Cliffs power plant

Albuquerque demo system

SG3 demo system

NH3 proof of concept

BGU dish

Start of Wizard Power

Start of Photovoltaic R&D

(with significant milestones ever since)

Start of Solar Thermal R&D

1970   1980   1990   2000   2010   2020

Appendix 4. Solar energy: Explaining the science

Summary

Solar energy is special. It is vast, ubiquitous and indefinitely sustainable. The solar resource is hundreds of times larger than all other available energy resources combined.

Solar energy utilises only very common materials; has minimal security and military risks; is available nearly everywhere in vast quantities; and has minimal environmental impact over unlimited time scales. No other energy source can make claims that come anywhere near these. Solar energy is a complete and long-term sustainable solution. Australia receives 30,000 times more solar energy each year than all fossil fuel use combined. Australia has a significant presence in the worldwide solar energy industry, which can be built upon to create a major export-oriented, technology-rich industry.

Energy supply options

There are five potentially available energy sources: the sun (in its various forms), nuclear energy (fission and fusion), fossil energy (coal, oil and gas), tidal energy and geothermal energy.

Solar energy is available on a massive scale, and is inexhaustible. It includes both direct radiation (photovoltaics and solar heat) and indirect forms such as biomass, wind, hydro, ocean thermal and waves. The direct solar energy resource is far larger than the indirect solar energy resource. Collection and conversion entails few environmental problems. Mass deployment entails minimal security risks, because of the intrinsic safety and wide distribution of the collectors. The ubiquity of solar energy precludes human conflict over access to it. Solar energy technology has minimal utility to regular armies or terrorists.

Nuclear energy from fission of heavy metals has substantial problems relating to nuclear accidents, nuclear weapons proliferation, nuclear terrorism, uranium and thorium deposit limitations, and waste disposal. Nuclear energy from fusion of light elements (similar to processes in the sun) is still several decades away from commercial utilisation, but may make a major contribution to sustainable energy supply in the future.

Fossil fuels are the principal cause of the enhanced greenhouse effect and are subject to resource depletion. Other problems include oil spills, oil-related warfare and pollution from acid rain, particulates and photochemical smog.

Tidal energy can be collected using what amounts to a coastal hydroelectric system. Geothermal energy in volcanic regions or from hot, dry rocks can be used to generate steam for district heating or to drive a steam turbine to produce electricity. Tidal and geothermal energy utilise relatively small and geographically restricted energy resources.

Photovoltaics

Photovoltaics (PV) is an elegant technology for the direct production of electricity from sunlight without moving parts. Most of the world's PV market is serviced by crystalline silicon solar cells. Sunlight causes electrons to become detached from their host silicon atoms. Near the upper surface is a 'one way membrane' called a pn-junction. When an electron crosses this junction it cannot easily return, causing a negative voltage to appear on the sunward surface (and a positive voltage on the rear surface). The sunward and rear surfaces can be connected together via an external circuit containing a battery or a load in order to extract current, voltage and power from the solar cell.

Hitherto, PV found widespread use in niche markets such as consumer electronics, remote area power supplies and satellites. Mass production is causing rapid reductions in cost. In recent years, the industry has expanded and costs have declined rapidly. PV systems are being installed on tens of millions of house roofs in cities, and also in large ground-mounted power stations. PV electricity is now less expensive than retail electricity from the grid throughout most of the world, and is approaching cost-competitiveness with wholesale conventional electricity. Over many decades, the cost of PV modules has been declining by about 18 per cent for each doubling of cumulative sales. This progress is likely to continue.

Most PV systems are mounted on fixed support structures. Some PV systems are mounted on sun-tracking systems to maximise output, while others use sun-tracking concentrators to concentrate light by 10–1000 times onto a small number of highly efficient solar cells. Hybrid PV/thermal micro concentrator systems are being developed to provide solar PV electricity, solar water heating, solar air heating, and solar air conditioning — a complete building energy solution.

PV systems mounted on house roofs can be used to achieve household carbon neutrality. A collector area of about 25 square metres is needed to carbon-neutralise a 5 star (energy rating) house with gas space heating, solar/gas water heating and efficient electrical appliances. Such a house exports more electricity to the grid during the day than it imports at night. An additional 5–10 square metres of PV panel is required to offset the annual greenhouse gas emissions of an efficient car.

PV panels on domestic and commercial building roofs compete with retail electricity prices, which are several times higher than wholesale electricity prices. The unsubsidised cost of rooftop PV generation has fallen below the daytime retail electricity price ('retail grid parity') throughout most of the world (except in northern latitudes). This is expected to drive rapid growth as hundreds of millions of home and commercial building owners adopt the technology. Grid parity has been achieved because of falling PV costs, rising fossil fuel costs, the introduction of carbon pricing and the introduction of time-of-use tariffs. Time-of-use tariffs properly reward PV systems for generating during sunny summer afternoons when peak loads caused by air conditioning, commerce and industry lead to high energy prices.

The efficiency of PV is eventually likely to rise above 60 per cent, compared with the current world record efficiency of 44 per cent. The cost of PV systems can be confidently expected to continue to decline for decades — as has happened with the related integrated circuit industry.

Solar thermal

Good building design, which allows the use of natural solar heat and light, together with good insulation, minimises the requirement for space heating. Solar water heaters are directly competitive with electricity or gas in most parts of the world. Solar air heaters will allow a large reduction in the heating load in many parts of the world, while solar driven air conditioning is a developing industry.

Solar thermal electricity technologies use sun-tracking mirrors to concentrate sunlight onto a receiver. The resulting heat is ultimately used to generate steam, which passes through a turbine to produce electricity. Concentrator methods are equally applicable to concentrating PV systems. The usual ways of concentrating sunlight are point focus concentrators (dishes), line focus concentrators (troughs, both reflective and refractive) and central receivers (heliostats and power towers).

Roof-mounted micro solar concentrators can harvest heat at temperatures of 100 to 300 degrees Celsius for use in hospitals, food processing, solar cooling and other light industrial applications, in competition with natural gas.

There is a large crossover between the technology of solar thermal and PV solar concentrators. The concentrating systems are quite similar, with the major technical difference being the solar receiver mounted at the focus: a black solar absorber in one case, and a PV array in the other. Since efficiencies are similar, the cost of energy produced by the two types of concentrator system is also similar.

An important future application of concentrated sunlight is the generation of thermochemicals and the storage of heat at high temperature in molten salt to allow for 24-hour power production. Concentrated solar energy can achieve the same temperatures as fossil and nuclear fuels, either directly (using mirrors) or through the use of chemicals (thermochemicals or bio fuels) created using concentrated solar energy. In the past, heavy industry (e.g., the steel industry) was often located near coalfields, in regions that are relatively poorly endowed with solar energy. Future steel mills could be built in the iron ore and solar energy rich Pilbara region of Western Australia.

Energy efficiency

Hand in hand with the utilisation of solar energy goes energy efficiency and conservation. 'Solar energy' and 'energy efficiency' is often the same thing. For example, an energy-efficient building is a building that utilises natural solar light and heat sensibly. Walking rather than driving a car uses a small amount of solar energy (food) rather than a larger amount of oil energy. A clothesline, solar salt production and putting on extra clothing displaces an electric clothes dryer, fossil-fuel fired kiln drying of salt and electric heating respectively.

Baseload power and storage

It is sometimes claimed, wrongly, that the absence of sunshine at night means that solar energy cannot dominate energy production.

Options for the provision of stable and continuous solar power include actively shifting loads from night to daytime; wide geographical dispersion of solar collectors to minimise the effect of cloud; precisely predicting solar energy output using satellite imagery and other detectors; diversification of energy supply to include all renewables; and energy storage.

Pumped hydro (whereby water is pumped uphill during the day and released through turbines at night to provide energy) is an efficient, economical and commercially available storage option that constitutes 99 per cent of current storage for the electricity industry. Lakes covering only 100 square kilometres (about five square metres per citizen), utilising either fresh water or seawater, would be sufficient to provide 24-hour storage of Australia's entire electricity production. Storage of heat from solar thermal electric systems in the form of molten salt and other media are attractive for spreading the production of electricity into the evening. Another future large-scale, day-night storage option is the batteries of millions of electric cars. In the longer term, long-distance high voltage DC transmission will further improve the robustness of a renewable electricity system.

Environmental impacts

The solar energy industry has minimal environmental impact. About 0.1 per cent of the world's land area would be required to supply all of the world's electricity requirements from solar energy. Indeed, the area of roof is sufficient to provide all of Australia's electricity, using PV panels.

We can never run out of the raw materials for solar energy systems because the principal elements required (silicon, oxygen, hydrogen, carbon, sodium and iron) are amongst the most abundant on earth. Old solar energy systems can be recycled without significant generation of toxic by-products. Gram for gram, advanced silicon solar cells produce the same amount of electricity over their lifetime as nuclear fuel rods. Per tonne of mined material, solar energy systems have 100-fold better lifetime energy yield than either nuclear or fossil energy systems.

The time that is required to generate enough electricity to displace the CO_2 equivalent to that invested in construction of a solar energy system is in the range of one to two years, compared with typical system lifetimes of 30 years. CO_2 payback and price are directly linked (via material consumption), and so CO_2 payback times will continue to fall, and will eventually decline to below one year.

The future of solar energy

Renewable energy technologies can eliminate the use of fossil fuels within a few decades, allowing a fully sustainable and zero carbon energy future.

Roof-mounted solar energy systems can provide PV electricity, hot water for domestic and industrial use, and thermal energy to heat and cool buildings and for steam production. Grid parity for PV at a retail level has already been achieved for most of the world's population. This is leading to rapid growth in sales in the residential and commercial sectors without the need for subsidies.

Large PV and solar concentrator power stations can provide most of the world's electricity. Concentration sunlight can provide process heat and thermochemicals.

Solar electricity, coupled with a shift to electrically powered cars and public transport, can provide most of the world's transport energy. A fleet of electric cars, each with large batteries, represents a large electricity storage facility to smooth supply and demand.

Direct competitiveness with fossil fuels for wholesale energy supply will be assisted by the implementation of full carbon pricing and the removal or equalisation of hidden support for fossil fuels.

In addition to direct solar energy collection, indirect forms of solar energy such as wind, biomass, wave and hydro are making important contributions. However, the indirect solar resource base is tiny in comparison with the direct sunshine utilised by PV and solar thermal, which will dominate a future all-renewable energy mix.

Whilst many technical adaptations to energy systems would be required to achieve a solar powered future, there are no insuperable obstacles. There are no significant environmental or material supply constraints. A switch to a zero carbon energy supply will cost considerably less than enduring severe climate change.

Appendix 5. Recommendations from the 1995 Expert Group on Renewable Energy Technologies

The Development and Use of Renewable Energy Technologies

Report of the Expert Group on Renewable Energy Technologies (EGRET) to the Minister for Primary Industries and Energy, February 1996

The previous chapter has considered many possible strategies that the Commonwealth Government could use to promote the development and use of renewable energy technologies. The Expert Group believes that to achieve a sustainable renewable energy industry, an integrated program of market and industry development is necessary. To enable the industry to take advantage of the opportunities that such a program will give it, it is important that the Commonwealth Government supports the industry with a consistent and long term program.

The Expert Group considers that the recommendations put forward will provide an effective and cost efficient framework of action for the Commonwealth Government to support the development and use of renewable energy technologies. Significant changes in energy use can be expected to take many years to achieve. The Expert Group therefore believes that it is essential that strong government action to support renewable should begin immediately. This will enable Australia to achieve the major social, economic and environmental benefits possible from increased use of renewable energy and a robust and internationally competitive renewable energy industry.

The Expert Group sees these recommendations as focusing on developing the opportunities for, and market strengths of renewable energy, as well as addressing the impediments to it.

The recommendations of the Expert Group are that:

- The Commonwealth Government declare the renewable industry as strategic for Australia and a key target for development assistance, because of its combination of environmental benefits, local and export market potential, local technological base, potential for industry growth and consequent future contribution to the economy.

- A peak renewable energy industry body be formed. The Commonwealth Government should provide $250,000 per annum for five years to this body for interface with Government on policy issues as they impact on renewable energy and for access to and implementation of renewable energy programs. Specific tasks for the peak body should include:
  - Investigating the merits of an export credit scheme for renewable energy technologies;
  - Undertaking a study of renewable energy support schemes, both in Australia and internationally;
  - Supporting an officer to focus on overseas markets and marketing, including seeking access by the Australian renewable energy industry to international development funds (additional $100,000 per annum from the Commonwealth to be matched by industry).

- A Standing Committee of the Australian and New Zealand Minerals and Energy Council (ANZMEC) be established on renewable energy, to report to the Council of Australian Governments (CoAG). This Committee would provide advice on the implementation of established policies as they impact on the renewable energy industry and also on further policy development. The Committee will consult with the peak industry body on relevant matters.

- The Commonwealth Government encourage through consultative mechanisms, such as the Council of Australian Governments (CoAG), the continuation of the reforms in the energy supply industries to ensure their operation on a fully commercial basis, including:
  - The removal, within 10 year, of implicit and explicit subsidies and cross-subsidies in energy pricing;
  - Mandating that energy service providers conduct least cost planning for all grid extensions, service upgrades, or substantial reconstruction and new service provisions, with the least cost alternative to be supplied to customers;
  - In recognition that the impact on the renewable energy industry resulting from these distortions is likely to continue for some time, governments providing compensating assistance to the renewable energy industry during the period of transition;
  - The delivery of community service obligations by means other than energy pricing.

- The Commonwealth Government establish explicit goals for 2000, 2010 and 2020 for the increased contribution of renewable energy sources to the Australian energy supply mix. Implementation programs should be developed at local, State/Territory and Commonwealth Government levels. Progress towards goals should be reported by the ANZMEC Standing Committee on Renewable Energy to CoAG.

- The Commonwealth Government in conjunction with State Governments (possibly via CoAG or the proposed ANZMEC committee) should initiate processes of reform of Commonwealth, State and Territory legislation and local government regulations to remove any barriers to the use of renewable energy products and systems.

- The Commonwealth Government support continuing technological innovation by the renewable energy industry and researchers by:
  - Establishing a 200 per cent tax concession for renewable energy research and demonstration projects;
  - Coordinating Commonwealth Government mechanisms for direct support of innovation and research on renewable energy technologies, products and systems;
  - Continuing support for Australian participation in the International Energy Agency's renewable energy programs and initiatives;
  - Providing $15 million annually to ERDC to support innovation and research for renewable, particularly demonstration and commercialisation.

- The Commonwealth Government strengthen existing information networks to facilitate provision of information appropriate to the needs of users, especially in rural areas. The Commonwealth should provide up to $2 million per annum for this purpose and funds should be matched by State/Territory governments on a dollar-for-dollar basis. The program should be developed and operated in conjunction with the proposed industry peak body.

- The Department of Employment, Education and Training support and promote training and accreditation schemes for workers in the renewable energy industry as in the current TAFE programs in renewable energy in Queensland and Victoria.

- The Commonwealth Government implement for a five year period, a national rebate scheme for purchases of renewable energy equipment by individuals for domestic purposes. The implementation would include associated promotional activities.

- The Commonwealth Government support the development of 'green energy' investment funds and facilitate their access to existing government concessional finance programs.

- The Commonwealth Government support exports of Australian manufactured renewable energy equipment by:

- Providing addition funds to AusAID to support Australian renewable energy equipment manufacturers;

- Providing long term funding support for the International Centre for the Application of Solar Energy;

- Undertaking renewable energy trade missions to key markets, led by senior Federal Ministers;

- Providing support to AustEnergy to market more effectively the Australian renewable energy industry.

- As part of its leadership role and to encourage demand the Commonwealth Government should purchase renewable energy using part of the savings from increased energy efficiency in its buildings (ie. five per cent of the 1992–93 building energy consumption by 1998–99 and 10 per cent by 2003–04 to be devoted to renewable energy). The proposed ANZMEC Standing Committee on Renewable Energy should examine this scheme for introduction in other jurisdictions.

- The Commonwealth Government should ensure that renewable energy is taken into account in the development of other key strategies, for example, forestry, regional development and eco-tourism.

- The Australian Bureau of Statistics be required and resourced to collect data on renewable energy equipment production and use.

Appendix 6. Photos

1. White Cliffs solar thermal power station, aerial view

Source: Andreas Luzzl.

2. Dr Ken Inall in the control room at White Cliffs

Source: Roger Gammon.

3. Stephen Kaneff cleaning the mirrors

Source: Roger Gammon.

4. White Cliffs

Source: Ray Dicker.

5. White Cliffs

Source: Roger Gammon.

6. Professor Stephen Kaneff at the plaquing of the White Cliffs site in 2006

Source: Roger Gammon.

7. White Cliffs solar pioneers; Kaneff, Wellings and Gammon at the plaquing of the White Cliffs site in 2006

Source: Roger Gammon.

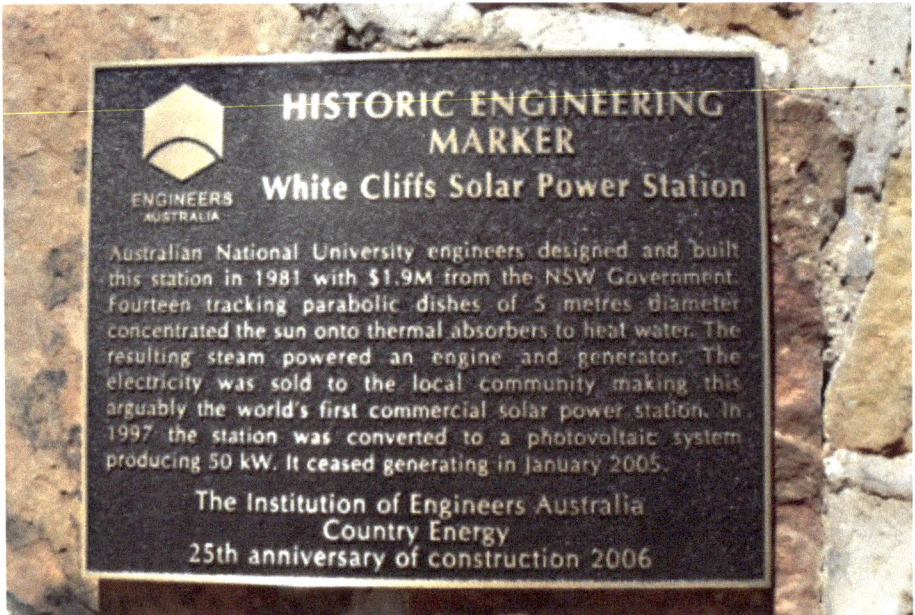

8. The 2006 Historic Engineering Site plaque

Source: Ray Dicker.

9. Professor Stephen Kaneff and Dr Keith Garzoli at Charlotte Pass

Source: Andrew Blakers.

10. 20 square metre dish for ammonia dissociation at ANU

Source: Andreas Luzzi.

11. 20 square metre dish at ANU

Source: Andreas Luzzi.

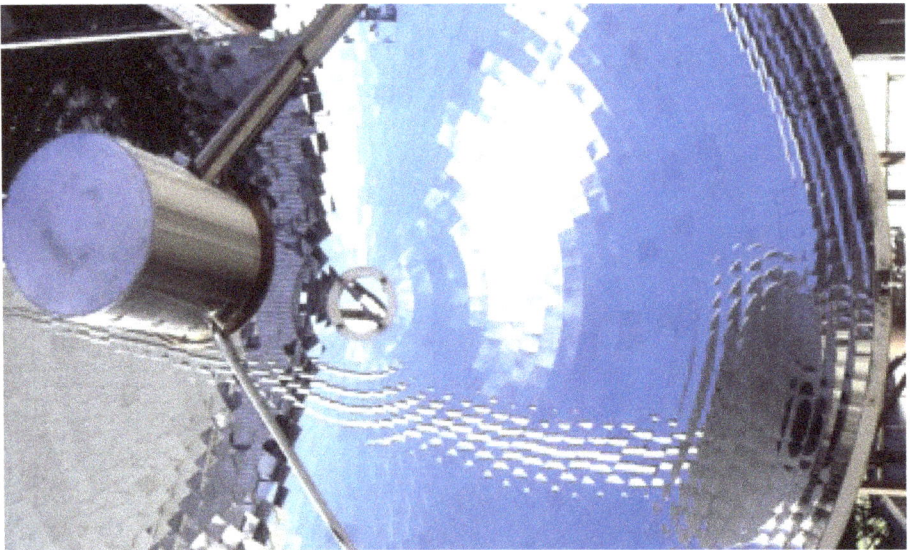

12. 20 square metre dish close-up

Source: Andreas Luzzi.

13. SG3: the 400 square metre "Big Dish" at ANU

Source: Andreas Luzzi.

14. 300 square metre Combined Heat and Powers System: Bruce College, ANU

Source: The Australian National University.

15. ANU PV-trough collectors at Rockingham, WA

Source: The Australian National University.

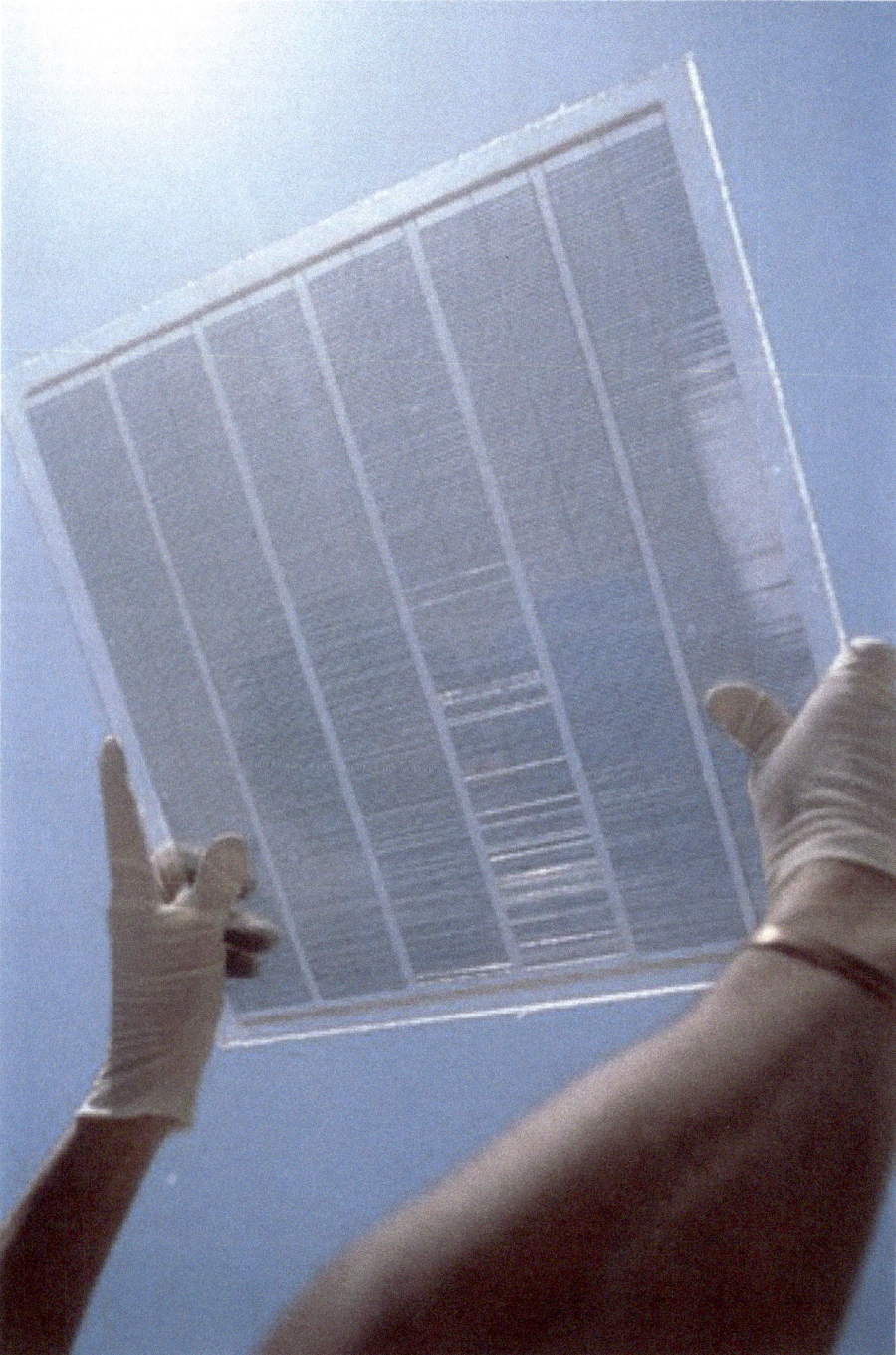

16. SLIVER solar module

Source: The Australian National University.

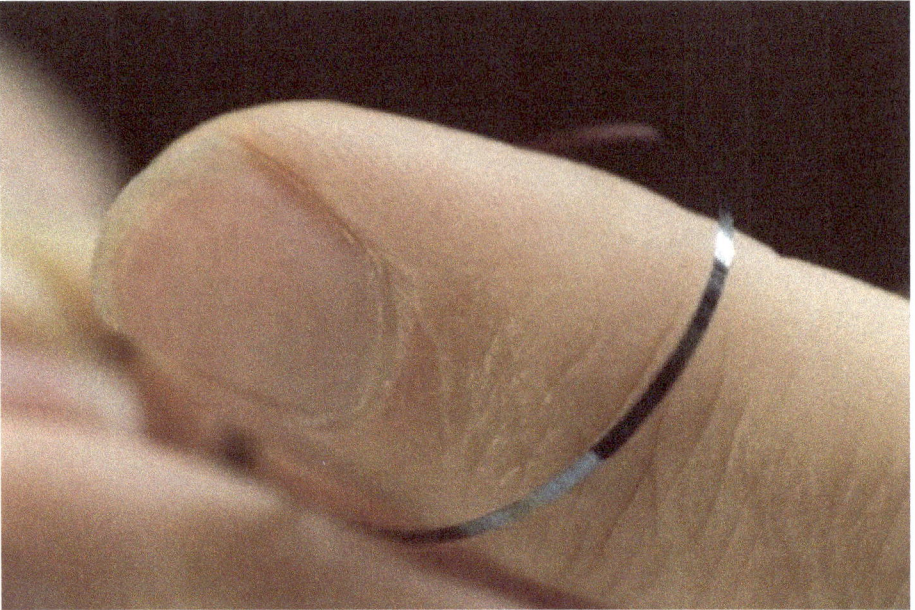

17. Flexible SLIVER solar cell

Source: The Australian National University.

References

ABC 2009, *Talking Heads*, http://www.abc.net.au/tv/talkingheads/txt/s2505093.htm

Abelson, P. and D. Chung 2004, *Housing prices in Australia 1970 to 2003*, Macquarie University Economics Research Paper, No 9/2004, September, Sydney.

ANUTECH Update, newsletter series, 1990–1999, Canberra.

Australian Bureau of Agricultural and Resource Economics 2007, *Australian energy: National and state projections to 2029–30: ABARE Research Report 07.24*, Canberra.

Australian Bureau of Agricultural and Resource Economics and Sciences 2011, *Energy in Australia*, Canberra.

Australian Bureau of Statistics (ABS), 1971a, 'Average weekly earnings, March quarter, 1971', http://www.ausstats.abs.gov.au/ausstats/free.nsf/0/CD0D874 A216BB4B6CA257516001098A3/$File/63020_MAR1971.pdf

—— 1971b, *Year Book Australia 1971*, http://www.abs.gov.au/AUSSTATS/ abs@.nsf/DetailsPage/1301.01971?OpenDocument

—— 1988, *People and dwellings in legal local government areas, statistical local areas and urban centres/(rural) localities, New South Wales*, Catalogue No. 2462.0 http://www.ausstats.abs.gov.au/

Australian Greenhouse Office 2003, *Renewable energy commercialisation in Australia*, Canberra.

Australian Nuclear Science and Technology Organisation (accessed 19 Aug 2009), http://www.ansto.gov.au

Ballinger, J. 1985, *The initial analysis of the Bonnyrigg Solar Village: School of*

Architecture research paper No. 6, Faculty of Architecture, University of New South Wales, Kensington.

Baverstock G. and A. Gaynor 2010, *Fifty years of solar energy in Australia and New Zealand: A history of the Australian and New Zealand Solar Energy Society*, http://www.auses.org.au/wp-content/uploads/2011/03/ANZSES-History.pdf

Birkeland, J. 2002, *Design for sustainability: A sourcebook of integrated, eco-logical solutions*, Earthscan, London.

—— 2008a 'Positive development', *Solar Progress*, August 2008, pp. 25–27.

—— 2008b *Positive development: From vicious circles to virtuous cycles*, Earthscan, London.

Birrell, N 2008, 'Science, innovation and how to get rich', http://www.monash. edu.au/research/swc/pac-science-journal.pdf

Bresnahan, T. and M. Trajtenberg 1995, 'General purpose technologies: engines of growth?', *Journal of Econometrics*, vol. 65, no. 1, pp. 83–108.

Caldicott, H. 2006, *Nuclear power is not the answer: To global warming or anything else*, Melbourne University Press.

Department of Primary Industries and Energy 1988, *Energy 2000: A national energy policy paper*, Canberra.

Diesendorf, M. 1992, 'Renewable energy: Overcoming the barriers', report for the Australian Conservation Foundation, 28 July, ARBN 007 498482.

Eberhart, M. 2007, *Feeding the fire: The lost history and uncertain future of mankind's energy addiction*, Harmony Books, New York.

Energetics Pty Ltd 1991, *Electricity end-use and CO2 emissions: Opportunities for change*, New South Wales Department of Minerals and Energy, Sydney.

Foster, S. and M. Varghese 1996, *The making of the Australian National University 1946–1996*, Allen & Unwin, Sydney.

Gammon, R. 1981, 'The solar energy programme in New South Wales', paper presented at the International Solar Energy Society (Australian and New Zealand Section) Conference, Macquarie University, Sydney, November.

Garnaut, R. 2008, *Garnaut Climate Change Review*, http://www.garnautreview. org.au/index.htm

Gilchrist, G. 1994, *The big switch: Clean energy for the twenty-first century*, Allen & Unwin, Sydney.

Hagen, D.L. and S. Kaneff 1991, *Application of solar thermal technologies in reducing greenhouse gas emissions*, report for the Department of the Arts, Sport, the Environment, Tourism and Territories, Commonwealth of Australia, ANUTECH Pty Ltd, Canberra.

Kaneff, S. 1991, *White Cliffs project: Overview for the period 1979–1989*, Office of Energy, Sydney.

Lowe, I. 1988, 'The energy policy implications of climate change', *Greenhouse: Planning for climate change*, CSIRO, Melbourne, pp. 602–12.

Meadows, D.H., D.L. Meadows, J. Randers & W.W. Behrens 1972, *The limits to growth: A report for the Club of Rome's project on the predicament of mankind*, Earth Island, London.

Millar, D.D. 1987, *The Messel era: The story of the School of Physics and its science foundation within The University of Sydney, Australia 1952–1987*, Pergamon, Sydney.

Morphett, J. 1990, 'Kick off', *ANUTECH Update*, 1, 1990. NSW Institute of Technology 1974, *Communique*. Nuclear Energy Institute (accessed 19 Aug 2009), http://www.nei.org

Ophel, T. and J. Jenkin 1996, *'Fire in the belly': The first fifty years of the pioneer School at the ANU*, Research School of Physical Sciences & Engineering, Institute of Advanced Studies, The Australian National University, Canberra.

Rosenthal, S. and P. Russ 1988, *The politics of power: Inside Australia's electric utilities*, Melbourne University Press.

United Nations Development Programme, Millennium Development Goals, Goal 7.2 (accessed 5 Sept 2009), http://www.undp.org/mdg/goal7.shtml

United Nations Framework Convention on Climate Change (accessed 29 Aug 2009), http://unfccc.int/2860.php

Vernon, K. 1976, 'An interpretation' in K. Vernon (ed.), *The oil crisis*, W.H. Norton, New York.

Warhurst, J. and A. Parkin 2000, *The machine: Labor confronts the future*, Allen & Unwin, Sydney.

Wing, P. 1991, *This gown for hire: A history of the Australian Tertiary Institutions Commercial Companies Association*, ANUTECH Pty Ltd, Canberra.

www.ingramcontent.com/pod-product-compliance
Lightning Source LLC
Chambersburg PA
CBHW061222270326
41927CB00022B/3464